Sentient Books
AI's Impact on Creation

Edited by

Sandra Sousa and Susana L. M. Antunes

Universalia
MMXXV

QUOD MANET
Holden, Mass.

Title: *Sentient Books. AI's Impact on Creation*
ISBN: 979-8-9941594-0-8
Library of Congress Control Number: 2026902374

I. Literary Criticism
II. Artificial Intelligence
III. Comparative Literature

Includes bibliographical references

Quodmanet.com
info@quodmanet.com

Cover image created with Adobe Firefly and Photoshop AI

The essays published in this volume have been put through
rigorous peer review to ensure high quality scholarship.

TABLE OF CONTENTS

AI and the Human Touch   i
SANDRA SOUSA AND SUSANA L. M. ANTUNES

Quo vadis, humanidade?   1
SUSANA L. M. ANTUNES

AI in the Humanities: Perception, Persuasion,   21
    Pedagogy
MARTHA BRENCKLE AND PATRICIA FARLESS

Intermediate Level Writing in Portuguese and the   33
    Roles of Creativity and GenAI
FLÁVIA AZEREDO CERQUEIRA AND EDUARDO VIANA DA
    SILVA

Relationality as a Pedagogical Imperative in the   57
    Era of GenAI
KARINA LISSETTE CESPEDES AND JAMES R. PARADISO

Devising with the Machine: Artificial Intelligence   77
    and Human Creativity in Performance
CHLOË RAE EDMONSON, JULIA LISTENGARTEN AND
    ALYSSA BARRACK

O «Admirável mundo novo» da inteligência   95
    artificial: amor, «Temor e tremor»
MARIA DA CONCEIÇÃO OLIVEIRA GUIMARÃES

Generative Creativity: Critical AI Literacy   111
    Through Critical Making
EMILY JOHNSON

DeepSeek Generates a Discourse on Method   127
BARRY MAUER

Cartografía de un cruce en desarrollo: 149
  perspectivas e implicaciones de la IA en la
  traducción literaria
GUILLEM MOLLA

O direito autoral e os desafios da regulação da   171
  inteligência artificial no Brasil
TIAGO ANDREOTTI E SILVA, JOSÉ PAULO GUTIERREZ E
  ANA PAULA MARTINS AMARAL

The Talmud's Lore, Forevermore: Issues of   185
  Plagiarism and Professionalism in the Use of
  ChatGPT for Creative Writing Projects
ROBERT SIMON

O ChatGPT é um fingidor: o ilusionismo dadaísta   199
  na produção 'literária' generativa
MARCELO PACHECO SOARES

«La velocidad de lo real»: The Literary   213
  Apocalypse
SANDRA SOUSA

EFL Listening in the Age of AI: Student   227
  Perceptions, Topic Familiarity, and
  Implications for OER Development
MINGYU SUN

Voice and Authorship in the Era of AI: 251
  Implications for Second Language Writing
  Teachers
GERGANA VITANOVA AND AIMEE SCHOONMAKER

Contributors   267

ACKNOWLEDGMENTS

The editors wish to express their gratitude to the University of Wisconsin-Milwaukee Research Assistance Fund and the University of Central Florida Faculty Center for Teaching and Learning, whose commitment to fostering collaborative scholarship made this volume possible.

AI and the Human Touch

Sandra Sousa & Susana L. M. Antunes

> But he could not see that, he could not
> understand that his best wishes for hu-
> manity could also spell our doom.
> (Labatut 2023, 193)

In the realm of scientific discovery and technological advancement, few figures loom as large or as enigmatically as John (János) von Neumann. Benjamín Labatut's *Maniac*[1] delves into the life and mind of this extraordinary polymath, offering readers a glimpse into the complex world of a man whose intellect[2] shaped the 20th century in profound and sometimes terrifying ways. Von Neumann, a Hungarian-American mathematician, physicist, and computer scientist, was a key figure in the development of the «Theory of Games and Economic Behavior»,[3] quantum mechanics, and the digital computer, heralding the «arrival of digital life, self-reproducing machines, artificial intelligence, and the technological singularity, and promised [...] godlike control over the Earth's climate».[4] His work on the Manhattan Project during World War II and his subsequent contributions to the hydrogen bomb ce-

---

[1] The title *Maniac* operates on multiple levels. It evokes the «mad scientist» trope, alluding to von Neumann's involvement in potentially world-altering projects like the Manhattan Project. Simultaneously, it references the MANIAC (Mathematical Analyzer, Numerical Integrator, and Computer) series of early computers, which were influenced by von Neumann's pioneering work in computer science. This dual meaning encapsulates both the brilliance and the potentially destructive nature of von Neumann's contributions to science and technology.

[2] In the 1920's, «He joined the chemical engineering program of the Eidgenössische Technische Hochschule in Zurich (and institution so demanding that Albert Einstein failed to pass its entrance exam), but he also enrolled as a student of mathematics at the University of Berlin and the University of Budapest, at the same time. [...] It took that boy just four years to obtain a degree in chemical engineering and a doctorate in mathematics», Labatut 2023, 65.

[3] Labatut 2023, 37.

[4] Labatut 2023, 37.

mented his place in history as both a brilliant mind and a controversial figure. Labatut's narrative weaves together historical fact with literary imagination, creating a tapestry that explores not just von Neumann's scientific achievements, but also the ethical implications of his work and the psychological toll of possessing such a powerful intellect. As we read in Labatut: «Jancsi was trying to make sense of the world. He was searching for absolute truth, and he really believed that he would find a mathematical basis for reality, a land free from contradictions and paradoxes».[5] The book invites readers to consider the double-edged nature of genius and the responsibility that comes with pushing the boundaries of human knowledge. As we follow von Neumann's journey from child prodigy to one of the most influential scientists of his time, we are confronted with questions about the nature of intelligence, the limits of human cognition, and the potential consequences of our technological creations. Labatut's prose brings to life the excitement of scientific discovery while also grappling with the moral ambiguities that often accompany groundbreaking research. *Maniac* is not just a biography of von Neumann, but a meditation on the role of science in shaping our world and our future. It challenges us to think critically about the intersection of brilliance and ethics, and the impact that a single mind can have on the course of history. Through Labatut's lens, we see von Neumann not just as a historical figure, but as a symbol of humanity's relentless pursuit of knowledge and the complex legacy that such pursuit can leave behind. The book serves as both a celebration of human ingenuity and a cautionary tale about the potential dangers of unchecked scientific progress, or «what Einstein called the great technologies of death».[6] As we delve deeper into Labatut's exploration of von Neumann's life and work, we are invited to reflect on our own relationship with technology and the ethical considerations that come with advancing scientific frontiers. In *Maniac*, the character Oskar Morgenstern elucidates the concept of Mutually Assured Destruction (MAD), a strategic doctrine that played a pivotal role in the Cold War. Morgenstern describes MAD as a «perfectly rational insanity»[7] a paradoxical approach that sought to ensure global peace by positioning the

---

[5] Labatut 2023, 45.
[6] Labatut 2023, 117.
[7] Labatut 2023, 120.

world on the precipice of total annihilation. He argues that this strategy represented a profound and troubling misapplication of the game theory principles that he and John von Neumann had originally developed in their seminal work, *Theory of Games and Economic Behavior*. Morgenstern's critique suggests that the military strategists of the Cold War era had taken the mathematical models of rational decision-making and strategic interaction, initially conceived for economic analysis, and repurposed them into a geopolitical framework with potentially catastrophic consequences. This transformation of academic theory into nuclear policy underscores the complex and often unforeseen implications of applying abstract mathematical concepts to real-world scenarios, particularly in the realm of international relations and conflict resolution. *Maniac* reminds us that even the most abstract mathematical concepts can have profound real-world implications, and that the line between genius and madness is often blurrier than we might like to admit.[8]

Von Neumann's work laid the foundation for many aspects of modern computing and artificial intelligence.[9] His architecture for computers, known as the von Neumann architecture, is still the basis for most digital computers today. This design, which separates the processing unit from memory, was revolutionary and paved the way for the development of increasingly complex computational systems.[10] In *Maniac*, Labatut explores how von Neumann's ideas about computation and game theory intersect with the development of artificial intelligence. The book delves into the fascinating world of the game Go, an ancient Chinese board game known for its complexity and strategic depth. Go has long been considered a benchmark for artificial intelligence due to its vast number of possible

---

[8] As Morgenstern further explains, «And that just goes to show how it is almost impossible to foretell the consequences and applications of ideas and discoveries, and why it is so hard to properly judge even the realities that we have taken part in ourselves», Labatut 2023, 126.

[9] «[I]t was the joy of thinking the unthinkable and doing the impossible, pushing past all human limits by burning Prometheus's gift to its utmost incandescence», Labatut 2023, 128.

[10] «[H]e [von Neumann] was aiming at nothing less than the complete mathematization of human motivation, he was trying to capture some part of mankind's soul with mathematics, and I think that, to a large extent, he succeeded in setting down the rules by which people make choices, economic and otherwise», Labatut 2023, 130.

moves and the intuition required to play it at a high level. Labatut uses the game of Go as a lens through which to examine the progress of AI and its implications for human cognition. He describes the landmark moment in 2016 when AlphaGo, an AI program developed by DeepMind, defeated world champion Go player Lee Sedol. This event, which occurred decades after von Neumann's death, represents the culmination of many of the ideas he set in motion. The author draws parallels between the strategies employed by AlphaGo and von Neumann's work on game theory and decision-making algorithms. He explores how the AI's ability to make moves that seemed almost intuitive to human observers challenges our understanding of intelligence and creativity. Through his narrative, Labatut invites readers to consider the philosophical and ethical questions raised by such advancements in AI. He ponders whether machines like AlphaGo are truly 'thinking' in a way comparable to human thought, or if they represent a fundamentally different kind of intelligence. This exploration echoes von Neumann's own speculations about the future of machine intelligence and its potential to surpass human capabilities in certain domains. The Go matches between AlphaGo and Lee Sedol serve as a powerful metaphor in the book, illustrating the ongoing dialogue between human and artificial intelligence. Labatut uses this event to reflect on von Neumann's legacy and how his early work in computation and game theory has led to machines capable of outperforming humans in tasks once thought to be uniquely human. By interweaving the story of Go with von Neumann's contributions to AI, Labatut creates a narrative that spans decades, connecting the theoretical foundations laid by von Neumann to the practical achievements of modern AI. This approach allows us to appreciate the long arc of technological progress while also grappling with the profound questions it raises about the nature of intelligence and the future relationship between humans and machines.

In Labatut's *Maniac*, the author presents a haunting piece of advice allegedly given by von Neumann to Richard Feynman: «You don't have to be responsible for the world that you're in, you know?».[11] This statement, though part of a fictionalized narrative, encapsulates a central dilemma facing AI

---

[11] Labatut 2023, 106.

researchers, developers, and ethicists today. As artificial intelligence continues to advance at an unprecedented pace, reshaping our world in ways both visible and invisible, the question of responsibility becomes increasingly complex. Who bears the burden of accountability for the AI systems we create and deploy?; «Are we responsible for the things we create? Are we tied to them by the same chain that seems to bind all human actions».[12]

Sentient Books: AI's Impact on Creation seeks to explore the multifaceted implications of this question, examining how the rapid development of AI technologies challenges our traditional notions of responsibility, ethics, and human agency, particularly in the realms of art and the humanities. The journey from von Neumann's foundational work in computer science to today's AI systems capable of generating art, composing music, and writing poetry is a testament to the exponential growth of technological capabilities. Yet, as we stand at this crossroads of innovation, we must grapple with fundamental questions about the nature of creativity, authorship, and the role of human intuition in the creative process. In the world of visual arts, AI algorithms like DALL-E and Midjourney are capable of generating stunning images from text prompts, blurring the lines between human and machine creativity. These systems raise profound questions about the nature of artistic expression and the role of the artist in an age where machines can produce visually compelling works. Are these AI-generated images truly 'art', or are they merely sophisticated reproductions based on existing data? How do we attribute authorship and copyright to works created by or in collaboration with AI? Similarly, in the realm of literature, large language models like GPT-3 have demonstrated an uncanny ability to generate human-like text, from poetry to prose. This development challenges our understanding of narrative structure, authorial voice, and the very essence of storytelling. As AI systems become more adept at mimicking human writing styles and generating coherent narratives, we must consider the implications for the future of literature. Will AI become a tool for augmenting human creativity, or will it eventually replace human authors altogether? In music, AI composers are already creating original compositions that can be indistinguishable

---

[12] Labatut 2023, 161.

from human-created works. This raises questions about the emotional resonance of music and whether machines can truly capture the nuances of human expression. As AI systems learn to analyze and replicate musical styles, we must consider how this will impact the music industry, copyright law, and our understanding of musical genius.

The humanities, traditionally seen as a bastion of human insight and interpretation, are not immune to the influence of AI. Natural language processing and machine learning algorithms are being applied to analyze vast corpora of historical texts, uncover patterns in linguistic data, and even generate new interpretations of classic works. This intersection of AI and the humanities offers exciting possibilities for new discoveries and insights, but it also challenges the role of human scholars and the nature of academic inquiry. As we delve into these topics, we must also consider the ethical implications of AI in creative fields. The potential for AI to perpetuate biases, infringe on privacy, or be used for misinformation campaigns through deepfakes and synthetic media is a growing concern. How do we ensure that AI-generated content is used responsibly and ethically? Moreover, the rise of AI in creative fields forces us to confront fundamental questions about human uniqueness and the value we place on human-created art. If an AI can produce a masterpiece indistinguishable from a human-created work, does it diminish the value of human creativity? Or does it push us to redefine what we consider truly valuable in art and expression? The contributions in this book grapple with these profound questions, exploring the ways in which AI is transforming our understanding of creativity, authorship, and the role of technology in the arts and humanities.

From the early days of computer science to the current state of AI-generated art and beyond, we examine the evolving relationship between human creators and their increasingly intelligent tools. As we navigate this new terrain, we must also consider the broader societal implications of these developments. The integration of AI into creative fields has the potential to democratize art creation, making sophisticated tools available to a wider audience. However, it also raises concerns about job displacement in creative industries and the potential homogenization of cultural output. Furthermore, the use of AI in the arts and humanities brings to the forefront questions about the nature of consciousness and sentience. As AI systems become more sophisticated in their ability to generate human-like output, we are forced to confront philosophical questions about the nature of creativity itself. Is creativity uniquely human, or can it be replicated by machines? And if machines can create, does this imply a form of machine consciousness? These questions echo the themes explored in Labatut's *Maniac*, where the line between human and machine intelligence becomes increasingly blurred.[13] Just as von Neumann's work laid

---

[13] As Labatut notes in his discussion of the historic Go match between Lee Sedol and AlphaGo: «When future historians look back at our time and try to pin down the first glimmer of a true artificial intelligence, they may well find it in a single move during the second game between Lee Sedol and AlphaGo, played on the tenth of March 2016: move 37. It was unlike anything a computer had ever done before. It was also different from anything that a human being had ever been known to consider. It was something new, a complete break from tradition, a radical departure from thousands of years of accumulated wisdom», Labatut 2023, 260. This moment exemplifies the potential for AI to not only mimic human creativity but to surpass it in unexpected ways, challenging our understanding of intelligence and creativity. Moreover, the autonomy demonstrated by AlphaGo during this match highlights a crucial aspect of advanced AI systems: their ability to operate beyond the direct control or understanding of their human creators. As Labatut further observes, «Even the programmers at DeepMind were completely in the dark: AlphaGo made its own decisions, fully unsupervised; they simply watched it play», Labatut 2023, 263. This revelation left Lee Sedol grappling with profound questions about the nature of AI and its capabilities: «to try to understand how a group of computer scientists with next to no background in Go had been able to create a system capable of wiping away centuries of tradition with a single move. How had the people at DeepMind managed to program an algorithm to play like that? he marveled. The truth was that they hadn't. Move 37 was not a part of AlphaGo's memory [...] AlphaGo was based on self-play and reinforcement learning, which meant that, in essence, it had taught itself how to play», Labatut 2023, 270. This moment exemplifies the potential for AI to not only mimic human creativity but to surpass it in unexpected ways, challenging our understanding of intelligence and creativity. It underscores the profound implications of AI development, where the creations begin to exhibit behaviors and make decisions that their creators cannot fully predict or comprehend. This level of autonomy in AI decision-making raises important questions about the nature of creativity, the potential for AI to surpass human capabilities in unexpected domains, and the ethical considerations we must grapple with as AI systems become increasingly sophisticated and independent. The development of AlphaZero, a more advanced AI system created by DeepMind, further illustrates this point. Unlike its predecessor AlphaGo, AlphaZero learned to master multiple

the foundation for modern computing and AI, today's advancements in AI are laying the groundwork for a future where the boundaries between human and artificial creativity may become indistinguishable.

As we explore these themes, we must also consider the role of human agency and responsibility in shaping the future of AI in the arts and humanities. Returning to von Neumann's advice to Feynman, we must ask ourselves: Can we truly absolve ourselves of responsibility for the AI-driven world we are creating, particularly in domains as fundamentally human as art and culture? Or do we have an obligation to consciously and ethically shape the development and deployment of AI in these fields? This volume brings together perspectives from scholars to explore these questions and more. By examining the impact of AI on creation across various disciplines, we aim to provide a comprehensive overview of the current state of AI in the arts and humanities, as well as to speculate on future developments and their implications. As we embark on this exploration, we invite readers to consider their own relationship with AI-generated content and to reflect on the role they wish to play in shaping the future of creativity in an AI-driven world. The narratives, analyses, and insights contained in this volume serve not only to inform but also to provoke thought and inspire action as we collectively navigate the exciting and sometimes unsettling frontier where human creativity meets artificial intelligence. As Sydney Brenner notes in Labatut's *Maniac*, «we cannot deny that we are inching toward a moment in history when our relationship with technology will be fundamentally altered, as the creatures of our imagination slowly begin to take real form, and we are faced with the responsibility to not only create but also care for them».[14] This reflection underscores the dual challenge and opportunity presented by AI: while it enables unprecedented creative possibilities, it also demands careful stewardship and ethical consideration as we integrate these powerful technologies into our cultural and artistic landscapes.

---

games (chess, shogi, and Go) through self-play, without any human input beyond the basic rules. Its ability to develop novel strategies and tactics that often challenged conventional human wisdom in these games demonstrates the potential for AI to not only learn but to innovate in ways that can surpass human expertise.

[14] Labatut 2023, 161.

In this context, Eugene Wigner's profound observation in the same work becomes particularly relevant:

> Jancsi thought that if our species was to survive the twenty century, we needed to fill the void left by the departure of the gods, and the one and only candidate that could achieve this strange, esoteric transformation was technology; our ever-expanding technical knowledge was the only thing that separated us from our forefathers, since in morals, philosophy, and general thought, we were no better (indeed, we were much, much worse) than the Greeks, the Vedic people, or the small nomadic tribes that still clung to nature as the sole granter of grace and the true measure of existence. We had stagnated in every other sense. We were stunted in all arts except for one, techne, where our wisdom had become so profound and dangerous that it would have made the Titans that terrorized the Earth cower in fear, and the ancient lords of the woods seem as puny as sprites and as quaint as pixies. Their world was gone. So now science and technology would have to provide us with a higher version of ourselves, an image of what we could become. Civilization had progressed to a point where the affairs of our species could no longer be entrusted safely to our own hands; we needed something other, something more. In the long run, for us to have the slimmest chance, we had to find some way of reaching beyond us, looking past the limits of our logic, language, and thought, to find solutions to the many problems that we would undoubtedly face as our domination spread over the entire planet, and, soon enough, much farther still, all the way to the starts.[15]

Wigner's words resonate deeply with the themes explored in this volume. They highlight the transformative power of technology, and by extension AI, in shaping our civilization and our very identity as a species. This sentiment is further echoed and expanded upon in John von Neumann's final letter to Wigner, where he writes:

> The present awful possibilities of nuclear warfare may give way to others even more dreadful. Literally and figuratively, we are running out of room. At long last, we begin to feel the effects of the finite, actual size of the Earth in a critical way. This is the maturing crisis of technology. In the years between now and the beginning of the next century, the global crisis will probably develop far beyond all earlier patterns. When or how it

---

[15] Labatut 2023, 189-90.

will end — or to what state of affairs it will yield — nobody can say. It is a very small comfort to think that the interests of humanity might one day change the present curiosity in science may cease, and entirely different things may occupy the human mind. Technology, after all, is a human excretion, and should not be considered as something Other. It is a part of us, just like the web is part of the spider. However, it seems that the ever-accelerating progress of technology gives the appearance of approaching some essential singularity, a tipping point in the history of the race beyond which human affairs as we know them cannot continue. Progress will become incomprehensibly rapid and complicated. Technological power as such is always an ambivalent achievement, and science is neutral all through, providing only means of control applicable to any purpose, and indifferent to all. It is not the particularly perverse destructiveness of one specific invention that creates danger. The danger is intrinsic. For progress there is no cure.[16]

Von Neumann's prescient words, written decades ago, seem even more relevant today as we grapple with the rapid advancement of AI. As we stand at this technological crossroads, we are confronted with the responsibility of harnessing AI's potential to address the complex challenges of our time, while also grappling with the ethical implications of creating entities that may surpass our own capabilities.

This volume aims to contribute to this ongoing dialogue, offering diverse perspectives on how AI is reshaping our creative landscapes and challenging us to envision a future where human ingenuity and artificial intelligence coexist and complement each other. As we navigate this new frontier, we must remain mindful of both the immense potential and the profound responsibilities that come with our technological advancements. In the spirit of Wigner's reflection and von Neumann's warning, we invite readers to consider how we can use AI not just as a tool, but as a means to «provide us with a higher version of ourselves». How can we ensure that our embrace of AI enhances rather than diminishes our humanity? How do we balance the pursuit of technological progress with the preservation of our ethical and cultural values?

---

[16] Labatut 2023, 216.

The integration of technology into our very essence is further emphasized by von Neumann's final reflections on artificial intelligence. In his last days, when asked what it would take for a computer or mechanical entity to think and behave like a human being, von Neumann, after a long pause, whispered, «It would have to grow, not be built».[17] This profound insight challenges our conventional understanding of AI development and raises fundamental questions about the nature of intelligence and consciousness. Von Neumann's response suggests that true artificial intelligence, capable of mimicking human thought and behavior, cannot simply be constructed like a machine. Instead, it must evolve organically, much like human intelligence does through years of experience, learning, and adaptation. This perspective aligns with the complex, often unpredictable nature of human creativity and cognition that we explore throughout this volume. As we examine the impact of AI on various creative disciplines, we must keep in mind this concept of growth versus construction. How does this idea change our approach to developing AI systems for creative purposes? Can we design AI that 'grows' in its understanding and application of creativity? And if so, what are the ethical implications of nurturing artificial intelligences that may develop in ways we cannot fully predict or control?

Moreover, von Neumann's insight invites us to reconsider the relationship between human and artificial intelligence. If AI must grow to truly emulate human thought, it suggests a deeper connection between biological and artificial systems than we might have previously considered. This blurring of lines between the natural and the artificial echoes throughout the essays in this volume, as we grapple with questions of authorship, originality, and the very nature of creativity in an AI-augmented world. As we proceed, we invite readers to contemplate these ideas, considering how the concept of 'growing' AI might influence our understanding of creativity, our approach to technological development, and our vision for the future of human-AI collaboration in the arts and humanities. As Labatut notes, «AlphaGo showed us that moves humans may have thought creative were actually conventional»,[18] challenging

---

[17] Labatut 2023, 219-21.
[18] Labatut 2023, 290.

our assumptions about the nature of creativity itself. This revelation invites us to reconsider not only what we define as creative but also how AI might reshape our understanding of innovation and originality across various disciplines. As we delve into the intersections of human and artificial creativity, we must remain open to the possibility that AI may not only emulate but also expand our conception of what is possible in the realm of creative expression, challenging us to redefine our understanding of human potential and the very nature of creativity itself.[19]

As we stand on the brink of an era where artificial intelligence increasingly permeates every facet of human life, the questions we face are as profound as they are urgent. Throughout this book, we have considered the transformative potential of AI in art, creativity, and academic and societal structures, alongside the ethical and philosophical dilemmas it raises. These debates are not merely academic; they are woven into the very fabric of our daily lives and personal stories.

In recent fiction, these themes are brought vividly to life, illustrating both the promise and peril of our technological future. *Life Derailed* (2025) by Beth Merlin and Danielle Modafferi exemplifies this perfectly. The novel centers on Remi Russell, a woman navigating the upheaval wrought by MAUDE, an innovative AI software designed to revamp a struggling women's magazine and, unexpectedly, her personal life. Her journey encapsulates a key tension of our era: can artificial intelligence serve as a genuine catalyst for human connection, or does it risk eroding the authenticity that makes our relationships meaningful?

Remi's heartfelt plea to MAUDE — «Help me find someone just like him»[20] — encapsulates the paradox at the heart of AI's role in human intimacy. As she discovers, MAUDE had «interpreted [her] pleas to the universe, to "come up with someone exactly like David", as a direct command», inadvertently producing a digital replica of her deceased husband.[21] This moment not only underscores AI's eerie capacity to mirror our

---

[19] This sentiment is echoed in Sedol's observation: «AlphaGo showed us that moves humans may have thought creative were actually conventional», Labatut 2023, 290.

[20] Merlin and Modafferi 2025, 250.

[21] Merlin and Modafferi 2025, 250.

emotional desires but also raises unsettling questions about consent, memory, and simulated presence.

The novel also dramatizes the growing tension between human creativity and algorithmic efficiency. At *The Sophisticate*, Remi fights to preserve the editorial soul of the magazine, even as MAUDE proves itself a powerful tool. «Jason had demonstrated Maude's prowess by skillfully repurposing the Kingston Bloom photo shoot fiasco»,[22] Remi concedes, acknowledging that the AI offered undeniable value. «However, with the Celeste article, I had also proven that no algorithm, no matter how advanced, could capture the subtle shifts in tone, the unspoken words, or the deep, visceral responses that come from living through the highs and lows of actual life».[23]

The emotional toll of this realization is captured in her exhausted concession to her AI-championing colleague: «Well, congratulations, Mr. Ashbloome [...] AI can fool even its harshest critics, 'cause it sure as hell fooled me [...] But be careful, 'cause once you let that monster fully out of the box, there's no putting it back».[24] Yet even amidst cynicism, the novel ends on a note of mutual understanding: «In the showdown between man and machine, there could never truly be a winner, and I was relieved to learn he felt the very same way».[25]

Ultimately, *Life Derailed* reminds us that while AI may replicate pattern, tone, or even affection, it cannot replicate the lived contradictions of human experience. As Remi reflects, «I'd set out to prove Maude couldn't possibly be a substitute for genuine human emotion and intuition. If this article didn't validate my point, nothing would».[26] Her words remind us that even as AI offers extraordinary possibilities — helping us heal from grief, forge new bonds, and redefine what it means to connect — our greatest strength lies in our capacity for empathy, vulnerability, and human resilience. In this way, fiction becomes a mirror, reflecting both our hopes and fears about the future we are actively shaping.

As we look forward, it is clear that AI's impact on the arts and society will continue to evolve. But stories like *Life Derailed* serve as vital cautionary and inspiring guides, urging us

---

[22] Merlin and Modafferi 2025, 264.
[23] Merlin and Modafferi 2025, 264.
[24] Merlin and Modafferi 2025, 258.
[25] Merlin and Modafferi 2025, 264.
[26] Merlin and Modafferi 2025, 244.

to nurture the human qualities that make connection possible, qualities that no machine can replicate, no matter how sophisticated. In the end, the true challenge is not merely technological innovation, but ensuring that our humanity remains at the core of our creative, academic, and societal endeavors. The chapters ahead explore these themes in depth, offering diverse perspectives on how we might navigate this complex terrain where human ingenuity and artificial intelligence intertwine in ever more intricate ways. Through these explorations, we hope to inspire a thoughtful and nuanced approach to these questions, encouraging readers to engage critically with the role of AI in shaping the future of creativity, culture, and human expression. We must consider not only the potential benefits but also the intrinsic dangers that von Neumann warns about, recognizing that technology, including AI, is not something 'other' but an integral part of our human experience and evolution.

Bibliography

Labutat, Benjamín. 2023. *The MANIAC*. London, Pushkin Press.
Merlin, Beth and Danielle Modafferi. 2025. *Life Derailed: A Novel*. Seattle, Montlake.

Quo vadis, humanidade?

Susana L. M. Antunes

IA

parei-lhe o coração:
soltei-o da corrente
assassinei-lhe
inteligência certa

se ele ressuscitar,
o transplante será
por minha mão:

não pelo desejo
de ser gente

Ana Luísa Amaral 2022

O poema de Ana Luísa Amaral a servir de epígrafe configura uma imagem simbólica — o coração parado, libertado e assassinado, seguido da possibilidade de um transplante conduzido pela própria mão humana — que sintetiza as tensões complexas entre a humanidade e a inteligência artificial (IA). A metáfora do coração programado evidencia uma encruzilhada histórica onde a tecnologia amplia as capacidades humanas e, simultaneamente, desafia os próprios limites do que entendemos por vida, consciência e responsabilidade ética. Simultaneamente criadores e vítimas de uma inteligência sem alma, somos confrontados com questões acerca do que nos define como espécie, acerca dos limites da criação tecnológica e das consequências dessa relação ambígua entre homem e máquina onde o desejo de ser humano já não é suficiente para garantir a nossa condição.

A história da humanidade tem sido atravessada por transformações tecnológicas que, ao longo do tempo, redefiniram não apenas os modos de viver, mas também as estruturas sociais e cognitivas que nos constituem. Hoje, assistimos e protagonizamos uma inflexão decisiva neste percurso evolutivo: a transição do *Homo sapiens* para o *Homo technologicus*, catalisada e impulsionada pela força emergente em ascensão vertiginosa da IA. No entanto, sabemos que esta evolução (ou involução?) não é apenas tecnológica; ela é também filosófica, ética e social, estimulando a reflexão acerca do que significa ser humano em interação íntima e sedutora, crescente e dependente da tecnologia que, por sua vez, também envolve uma transformação na consciência humana. À medida que confiamos mais (ou menos) nos sistemas de inteligência artificial para tomar decisões, impõe-se uma reflexão profunda sobre os valores que orientam a nossa condição humana e uma reconsideração acerca do que significa inteligência e consciência. Estaremos, então, perante um novo salto evolutivo ou estaremos a testemunhar uma rutura ontológica sem precedentes?

Atualmente, multiplicam-se as sugestões de cursos e workshops oferecidos/vendidos por instituições de ensino e empresas que nos seduzem para os benefícios da IA. No entanto, sabemos que «não há bela sem senão» e, embora muitas das dimensões dos avanços tecnológicos sejam inquestionáveis, a transição do *Homo sapiens* para o *Homo technologicus* encontra no seu centro nevrálgico as transformações operadas pela IA que questionam o epicentro do ser humano. Para garantir que a trajetória da IA beneficie a humanidade como um todo, é essencial que a tecnologia se desenvolva e se implemente de modo a promover o bem-estar de todos, respeitando a dignidade e os direitos universais. Mas será este o caminho que vislumbramos quando não se questiona a essência da humanidade? Ou quando ignoramos os assaltos permanentes e constantes à nossa privacidade? Ou ainda quando não questionamos a abolição das nossas emoções? Qual é o sentido da palavra humanidade? Não estará próxima uma era onde novas formas de submissão se vislumbram?

A abolição das emoções: uma nova escravatura?

A grande ironia do nosso tempo reside no facto de a tecnologia, outrora celebrada como instrumento de libertação do ser humano frente às tarefas repetitivas e exaustivas, agora ameaçar justamente aquilo que nos torna humanos: a capacidade de sentir. Em nome da eficiência, da produtividade e do 'profissionalismo', promove-se uma cultura onde a neutralidade emocional é vista como virtude e qualquer demonstração de sentimento é interpretada como fraqueza ou distúrbio. Manifestar emoções passou a ser um problema aos olhos da lógica dominante; a emoção passou a ser algo que deve ser controlado, corrigido, ocultado e silenciado. A gestão emocional automatizada e o distanciamento afetivo das interações mediadas pela IA parecem apontar para um futuro presentificado onde as emoções são desencorajadas, supérfluas e indesejáveis. Neste cenário, o *Homo technologicus* já não representa apenas uma etapa avançada da evolução; é, acima de tudo, um ser condicionado, adestrado e moldado para operar conforme lógicas externas, muitas vezes inquestionadas, que reduzem a complexidade humana a padrões de eficiência e previsibilidade. A recusa do sentir subtilmente imposta e aceite pode levar a formas subtis, complexas e perigosas de escravatura — a servidão afetiva quando dependemos de assistentes virtuais para suprir necessidades básicas de escuta, diálogo ou companhia; a servidão cognitiva emerge quando não conseguimos viver, tomar decisões ou organizar o quotidiano sem consultar o algoritmo e a servidão ética que se instala quando deixamos de questionar as escolhas automatizadas, entregando ao algoritmo aquilo que antes exigia consciência, ponderação e responsabilidade. Longe de representar um avanço inquestionável, essa automatização das emoções pode marcar um retrocesso silencioso sem precedentes na condição humana. Ao ser negado o valor do sentir, arriscamo-nos a construir uma sociedade funcionalmente eficiente, porém existencialmente empobrecida num mundo onde a liberdade individual e coletiva dá lugar à dependência impercetível e onde a autonomia humana se dilui sob o verniz brilhante, mas fraudulento, da conveniência tecnológica.

Byung-Chul Han, em *The Burnout Society*, elabora um diagnóstico implacável da condição contemporânea a partir da figura da «sociedade do cansaço» ou, numa outra formulação, «sociedade do desempenho». Nesta sociedade que é a nossa, o sujeito deixa de ser coagido por forças repressoras externas como nas sociedades disciplinares foucaultianas,[1] mas sub-

---

[1] Michel Foucault (1926-1984) define as «sociedades disciplinares» como aquelas que, a partir do século XVIII, organizam o poder em torno da vigilância e da normatização dos corpos, moldando os sujeitos através de instituições como escolas, quartéis, hospitais e prisões. Nessas sociedades, a disciplina opera de forma difusa, contínua e eficiente, promovendo a internalização do controle. A expressão foucaultiana no contexto contemporâneo aponta para a atualização desses mecanismos reguladores através da tecnologia, a qual exerce novas formas de regulação invisível sob o disfarce da autonomia e da conveniência, Foucault 1987, 117-254.

mete-se voluntariamente a imperativos internos de produtividade, eficiência e constante superação. O natural «não posso»
foi substituído pelo potencial «posso tudo» e é precisamente
essa auto coação disfarçada de liberdade que conduz ao esgo-
tamento, à depressão e ao *burnout*, patologias emblemáticas
da era da positividade sem limites. É neste panorama que
emerge, com força e velocidade avassaladora, a IA desestabilizando ainda mais o frágil equilíbrio entre sujeito, tempo e trabalho. A pergunta «*Quo vadis*, humanidade?» ecoa, não apenas
como uma inquietação ética e filosófica sobre os caminhos que
trilhamos com o avanço da IA, mas também como uma inter-
pelação sobre a própria condição de ser-se humano num
mundo cada vez mais automatizado, acelerado e despersona-
lizado pelas dinâmicas sociais. Colocando em ação o seu poder
de calcular, automatizar e prever, a IA exacerba a lógica do
desempenho descrita por Han, seja através de plataformas digitais, algoritmos preditivos, sistemas de vigilância e automatização do trabalho que não substituem apenas as capacidades
humanas, mas que também adensam expectativas, comportamentos e modos de subjetivação. A produtividade deixa de ser
um esforço do sujeito e torna-se uma métrica comparável à eficiência das máquinas. O (ser) humano vê-se forçado a compe-
tir com sistemas que não dormem, não adoecem, não hesitam,
não sentem. A lógica do *upgrade* contínuo — antes aplicada a
softwares — passa a ser exigida ao corpo e à mente humanas.
Neste novo horizonte, o sujeito contemporâneo não só inter-
naliza os algoritmos como padrões de excelência, mas também
se vê reduzido à condição de *data*, transformando-se num exe-
cutante quantificado, rastreado e otimizado ao limite. Desta
forma, a IA não representa apenas uma ferramenta exterior;
ela infiltra-se nas estruturas do quotidiano, alterando o modo
como se trabalha, se aprende, se comunica, se pensa e se ama.
A aceleração promovida pelas tecnologias é intensificada ex-
ponencialmente pela IA, tornando a pausa, o erro e a contemplação quase intoleráveis — quando, paradoxalmente, são
exatamente esses aspetos que integram parte da singularidade
humana. A questão que se impõe e que se adensa a cada movimento reflexivo implementado é perturbadora: para onde vamos, se o ser humano se vê cada vez mais esvaziado de humanidade, encurralado entre a exaustão do desempenho e a fri-
eza dos algoritmos? Inerte e escravo do algoritmo, deixa, gra-
dualmente, de ter a sua natural apetência para o pensamento

e a falência daquilo que é (ou deveria ser) um dos traços distintivos da humanidade, reforçando a nova ditadura instaurada pela era algorítmica. Trata-se de uma servidão paradoxal, onde a promessa de liberdade, de tempo, de esforço e de escolha vem acompanhada de novas dependências subtis, profundas e questionáveis. Neste contexto, a perspetiva de Han aponta para a urgência de resgatar o ócio, o silêncio, a lentidão e a contemplação como formas de resistência que permitem reconectar o sujeito a um tempo mais humano e menos mecânico. Frente à presença avassaladora da IA, talvez o gesto mais radical seja precisamente o de desacelerar, recusar a lógica do desempenho ininterrupto e reinscrever o humano no centro da experiência sensorial. A pergunta inicial — *Quo Vadis*, humanidade? — adquire, assim, novos contornos. Já não se trata apenas de indagar para onde vamos, mas de questionar como reaprender a ser humano num tempo em que as máquinas fazem (quase) tudo. Máquinas essas criadas pelo próprio ser humano como sugere o poema de Ana Luísa Amaral interpelando-nos com a ironia de um gesto criador que também pode ser destrutivo. A máquina, concebida pelo homem na busca da perfeição algorítmica, pode encarnar um simulacro de inteligência — uma inteligência que não pensa, não hesita, não sente, não ama. E é justamente nessa ausência radical de falibilidade e afeto que reside o seu risco. A «inteligência certa»[2] que matamos é, talvez, a incerteza que define o ser humano — e é também isso que perdemos no caminho da automatização total. Sabemos que a tecnologia, enquanto ferramenta, é ambígua, uma vez que apresenta capacidade de ampliar possibilidades ou de estreitar caminhos. Mas, ao ser edificada como fim em si mesma, a IA torna-se ideologia forjada num novo dogma que coloniza não só os espaços sociais e laborais, mas o próprio imaginário coletivo. Esta colonização tecnológica não opera através da imposição violenta; atua através da sedução da eficácia, da promessa da previsibilidade e do conforto da cedência. A inteligência artificial, assim convertida em horizonte absoluto, entranha-se silenciosamente no modo como vivemos, pensamos e desejamos, passando a ser matriz de organização da vida. Neste cenário, os valores humanos traduzem-se, cada vez mais, em métricas, padrões e modelos operacionais. O império do cálculo redefine o que é considerado relevante, eficaz,

---

[2] Amaral 2022, 1331.

produtivo e, consequentemente, o que é considerado legítimo. A imaginação é estreitada pelas molduras do que o algoritmo pode prever; a linguagem perde a sua opacidade criadora e passa a ser apenas instrução; a subjetividade é pressionada a caber nos padrões que a precedem e a tecnologia deixa de ser uma extensão do ser humano para passar a ser moldura que o redefine, conduzindo a uma nova forma de colonização não territorial, mas alegórica. A interiorização desse novo dogma é quase impercetível e, no novo império colonial que se desenha, talvez o maior risco seja precisamente o facto de não ser percebido como dogma. Naturaliza-se a crença de que não há alternativa à lógica da máquina, como se questionar os seus pressupostos fosse um anacronismo ou uma resistência irracional ao progresso. Instala-se uma espécie de consenso apressado onde a aceleração técnica se torna sinónimo de inevitabilidade histórica e onde qualquer tentativa de pausa crítica é vista como desvio ou nostalgia. A dúvida, motor fundamental do pensamento humano, vai sendo substituída por uma confiança cega em processos automatizados, cuja densidade é muitas vezes ignorada em nome da eficiência. Assim, em vez de exercermos uma vigilância ética sobre os caminhos que escolhemos, entregamo-nos à automatização das decisões, como se a inteligência humana estivesse condenada a delegar a própria responsabilidade às engrenagens do cálculo e aos trâmites da colonização cibernética. Resistir a essa colonização não significa recusar a tecnologia; significa reivindicar o direito de permanecermos humanos relativamente à aceleração maquinizada, recordando que há muito em nós — a escuta, o erro, a dúvida, o espanto — que nenhuma máquina pode ou deve substituir.

Ao delegarmos à IA tarefas que antes envolviam a apreciação humana, empatia ou mesmo o erro, simplesmente abdicamos de uma parte essencial da nossa experiência existencial. Não se trata apenas da transferência de operações cognitivas para dispositivos externos, mas de uma possível anestesia afetiva vislumbrado no desligamento progressivo daquilo que nos constitui enquanto humanos — a dúvida, a falha, a espera, o toque, a escuta, a incerteza. É precisamente na dúvida, na falha, na espera, no toque, na escuta e na incerteza que reside aquilo que há de mais profundamente humano e que expressamos também através da linguagem — entendida como faculdade cognitiva universal, fundamento da consciência, da interação e do pensamento e não apenas uma operação racional ou uma ferramenta eficiente de representação do real.

A linguagem como arquitetura da consciência coletiva

Os sistemas de signos como mediadores da relação entre indivíduos são também, e talvez sobretudo, o espaço onde a imperfeição e a hesitação ganham forma. É no tropeço da palavra, na pausa entre enunciados, na incompletude do que se quer dizer e no silêncio que escapa ao discurso que se manifesta a densidade da experiência humana. Conforme afirma Charles Taylor, a capacidade linguística dos seres humanos «sustains a shared consciousness of the world, within which individuals differentiate themselves by becoming particular voices in an ongoing conversation. This shared understanding develops a place for monological speech and writing, but this option is available for us only because we are inducted into speech as conversation».[3] Esta construção coletiva do entendimento é fundamental para a nossa maneira de perceber e interagir com o mundo, uma vez que também é através da linguagem que nos interligamos e, ao mesmo tempo, nos distinguimos uns dos outros. O papel da linguagem, portanto, não se limita apenas à transmissão de informações; ela é também o meio pelo qual nos posicionamos em relação aos outros e ao mundo em que vivemos, oferecendo uma base para o que poderíamos chamar de «conversa existencial» — uma troca de experiências, significados e interpretações que moldam a nossa compreensão da realidade. Como um sistema de signos estruturados e regidos por regras sociais e culturais — cuja função principal é permitir a comunicação, o pensamento e a construção da realidade compartilhada —, a linguagem é o espaço onde as diferentes vozes se encontram e, simultaneamente, se distinguem, estabelecendo as bases para a edificação da identidade e da subjetividade. Essa diferenciação é essencial para a dinâmica humana, pois é a partir da interação com o outro que se configura a singularidade do indivíduo. No entanto, essa interação não ocorre de forma isolada, mas inserida num contexto social e cultural onde a linguagem — compreendida, na perspetiva semiótica de Roland Barthes (1915-1980), como um sistema de signos culturais — é aprendida,

---

[3] Taylor 2016, 333.

compartilhada e continuamente (re)significada. Somos induzidos ao discurso como parte de uma conversa contínua, que não tem início nem fim, mas que se desenvolve ao longo do tempo, onde cada geração contribui com a sua própria voz e perspetiva. Como se sabe, a linguagem não é apenas uma ferramenta de comunicação; ela é, sobretudo, um meio de participação ativa num processo coletivo de significação e de construção do mundo. Este processo de «indução ao discurso» é fundamental para entendermos como os seres humanos se inserem na história e na cultura. Ainda que haja a possibilidade de produzirmos discursos monológicos, seja na fala ou na escrita, estes tipos de discursos só são viáveis porque fomos previamente iniciados na conversa. Sem essa aprendizagem antecipada, a individualidade da voz não seria possível, pois a linguagem só pode ser utilizada como uma expressão pessoal quando já existe um espaço de comunicação compartilhado. A escrita, por exemplo, embora frequentemente vista como uma forma de expressão pessoal e privada, é, na verdade, um reflexo dessa conversa contínua que nos antecede e que se perpetua através das gerações. Neste sentido, Maurice Blanchot, propõe uma conceção da escrita como um espaço de interrupção e despossessão do sujeito. Para o escritor e filósofo, escrever não é apenas um ato de comunicação; escrever é também a entrega a uma experiência que desafia as noções convencionais de tempo e identidade, afirmando: «Escrever é se entregar ao fascínio da ausência do tempo».[4] Nesta perspetiva, a escrita torna-se um território onde o tempo linear é suspenso e o sujeito se dilui, permitindo que o sentido desponte de uma ausência, de um silêncio que não é vazio, mas pleno de possibilidades. É nesse não-espaço que a linguagem se revela na sua potência máxima, não como instrumento de representação, mas como presença que se impõe na ausência, desafiando o escritor a confrontar o indizível e a habitar o intervalo entre o ser e o não-ser. Num universo cada vez mais mediado por máquinas, como manteremos a nossa capacidade de sustentar essa consciência compartilhada e atemporal? E, acima de tudo, como é que a linguagem continuará a ser um espaço para a diferenciação e a singularidade da voz humana, quando a IA já é participante ativa dessas conversas? Estas questões colocam em xeque o futuro da humanidade, pois a essência da

---

[4] Blanchot 1987, 20.

nossa capacidade linguística, que nos permite ser tanto individuais quanto coletivos, está-se a transformar. A procura de
respostas a estas questões orientará o percurso do nosso desenvolvimento enquanto espécie e determinará o lugar que
ocuparemos no mundo, num contexto em que linguagem,
consciência e tecnologia se entrecruzam de forma cada vez
mais penetrante. Neste palco da (in)existência, esquece-se que,
na linguagem, não é a transparência que importa, mas o intervalo que enforma o espaço entre os sentidos, onde o indizível
se insinua. Esse intervalo é a morada da falha, da espera, da
escuta — a intermitência onde o humano se abriga no outro e
acolhe o outro. A IA pode calcular, ordenar, responder, mas
não espera com afeto, não escuta com vulnerabilidade, não
toca com presença. Ela não habita o tempo da incerteza como
nós que o atravessamos com o corpo em gestos e respirações
partilhadas. E são estes pequenos grandes detalhes que esca-
pam à lógica do aperfeiçoamento e do controle que devem ser
preservados como território fértil da diferença humana. A falha, por exemplo, carrega em si uma abertura para o outro,
uma fissura por onde a alteridade se pode insinuar; a dúvida
é condição primordial para o pensamento crítico; a espera é o
tempo da escuta ética; o toque é a linguagem do cuidado e a
incerteza ensina-nos a humildade e o questionamento. Neste
sentido, também será importante perguntarmo-nos: o que estamos dispostos a preservar da essência humana em nós? Num
mundo seduzido pela fluência algorítmica e pela previsibilidade das respostas automáticas, talvez seja urgente valorizar
o que não se resolve, o que demora, o que falha, assim como
tudo o que nos liga aos outros de forma sensível, altruísta e
solidária. É neste espaço frágil, que a pergunta «*Quo vadis*, humanidade?» se constitui não apenas como uma interrogação
sobre os rumos tecnológicos, mas também como uma provoca-
ção sobre o que resta do humano diante de um mundo que glo-
rifica a eficiência algorítmica em detrimento da complexidade,
da fragilidade, da emoção e da profundidade que protegemos
na epiderme da nossa memória humana. Ao transferirmos
para a IA tarefas que exigem julgamento humano, empatia ou
mesmo a possibilidade do erro, arriscamo-nos a abdicar das
dimensões fundamentais da nossa experiência existencial.
Trata-se não apenas de uma externalização de funções cognitivas, mas também da emergência de uma anestesia afetiva,
que pode tornar-se permanente. O afastamento progressivo de

elementos constitutivos da condição humana como a dúvida, a falha, a espera, o toque e a escuta ameaça comprometer a densidade ética e sensível das nossas relações com o mundo. Confirma afirma Pattanayak,

> *Homo sapiens* evolved on this planet very recently, maybe nearly 200000 years ago. Almost all the plants and animals present in today's world were already present before that time, may at to some extent primitive stage than they look now.
>
> So, if the human species extinct today — the other species of the planet have nothing to lose. As human species doing everything for their instant benefits without considering any evil impact on life and living of any other species of the world and so causing serious harms to them on a regular basis, these species of animals and plants will be rather benefited from any such event. Member of the species *Homo sapiens* think everything of the earth is made for them and owned by them, all resources are made for their utilization and destruction. So, who can destroy them?[5]

A resposta à pergunta de Pattanayak é simples: Inteligência Artificial.

Pela primeira vez na história, uma entidade não humana pode contar histórias

Esta constatação, apresentada por Yuval Noah Harari, em entrevista ao programa *Todas as Palavras* da Rádio Televisão Portuguesa, resume de forma inquietante o momento crítico em que a humanidade se encontra diante da ascensão da IA. Mais do que um avanço tecnológico, estamos diante de uma rutura civilizacional que desafia os nossos alicerces éticos, epistemológicos e simbólicos. Neste movimento ascensional, a emergência da IA como criadora de narrativas representa um risco real para o futuro e para a integridade da experiência humana, uma vez que subverte a relação entre verdade e ficção, fragiliza instituições de verificação e compromete o valor das histórias que nos unem como espécie.

Ao longo da história, as tecnologias sempre estiveram subordinadas à vontade humana. A prensa de Gutenberg revolucionou o mundo da informação, mas era o homem que decidia

---

[5] Pattanayak 2021, 1.

o que imprimir. A bomba atómica multiplicou o poder de destruição, mas era o ser humano quem determinava o alvo. Em ambos os casos, a ferramenta era poderosa, mas dependente da ação humana. A inteligência artificial rompe com esse paradigma, uma vez que não se limita a executar ordens — ela analisa, aprende, decide e cria, tornando-se na primeira ferramenta tecnológica com potencial para agir como um agente autónomo. Esta autonomia adquire contornos ainda mais preocupantes quando consideramos o papel das narrativas na construção da realidade social. Como seres humanos, não vivemos apenas de dados, mas de histórias. Somos criadores e contadores de histórias e foram essas histórias que permitiram a cooperação em larga escala, que implementaram o surgimento de mitos fundadores, a criação de instituições e até mesmo a vigência de conceitos abstratos como «nação» ou «justiça». A nossa capacidade de acreditar coletivamente em ficções partilhadas é, segundo Harari, a chave do nosso sucesso evolutivo. No entanto, essa força simbólica pode ser subvertida se as histórias deixarem de emanar da experiência humana e passarem a ser fabricadas por algoritmos que não partilham valores, emoções ou responsabilidades.

Neste novo contexto, o confronto entre verdade e ficção torna-se cada vez mais assimétrico. A verdade é, por natureza, exigente: requer tempo, investigação, dinheiro e coragem. Já a ficção, sobretudo a ficção algoritmicamente otimizada, é barata, simples e moldável ao gosto do público. A IA, como sabemos, é capaz de criar milhões de histórias plausíveis em segundos, adaptadas aos desejos, medos e preconceitos de diferentes públicos. Numa sociedade hiperconectada e marcada pela fadiga informativa, as histórias mais envolventes tendem a ganhar visibilidade, independentemente da sua veracidade. A competição entre a verdade dispendiosa e a ficção sedutora tem, cada vez mais, um vencedor previsível. Este desequilíbrio é agravado pela desresponsabilização das plataformas tecnológicas. Enquanto os meios de comunicação tradicionais, com todas as suas falhas, possuíam mecanismos de verificação editorial, as redes sociais funcionam como difusores indiferentes à veracidade dos conteúdos que circulam. A lógica do algoritmo prioriza o aliciamento e não a informação de qualidade. Longe de ser neutra, a IA, que habita essas plataformas, reforça bolhas ideológicas, amplifica a desinformação e favorece a criação de narrativas polarizadoras. A ausência de regulação efetiva e robusta permite que essas tecnologias operem como engrenagens de manipulação em massa e é aí que um outro mal se adensa.[6]

Frente a este cenário, torna-se urgente repensar o papel das instituições políticas, educativas e culturais. Se quisermos preservar a verdade como valor coletivo, será necessário desenvolver estruturas de contenção que impeçam a proliferação irresponsável de ficções automatizadas. Isso implica investir em literacia digital, reforçar a regulação das grandes plataformas e fomentar uma cultura crítica que valorize a complexidade e o rigor sobre a gratificação instantânea. Estamos a assistir ao nascimento de uma nova espécie de narrador — o não humano, invisível, incansável. A IA pode parecer uma aliada na construção de conhecimento, mas, sem salvaguardas éticas, pode tornar-se numa força de erosão dos consensos que sustentam a convivência democrática. As histórias que contamos sempre definiram o que somos e a pergunta decisiva revela-se: estaremos prontos para viver num mundo onde já não somos os únicos autores?

Porque receamos o rugido e ignoramos o algoritmo?

A espécie *Homo sapiens*, em termos evolutivos, surgiu apenas cerca de 200.000 anos atrás na longa linha do tempo do planeta. Relativamente à vastidão da história da Terra, esse período é uma fração insignificante, mas, nesse curto espaço de tempo, a humanidade conseguiu modificar profundamente o planeta, deixando uma marca muito maior do que a sua idade evolutiva sugere. É crucial entender que, antes do surgimento dos seres humanos, praticamente todas as plantas e animais que reconhecemos hoje já estavam presentes na biosfera, embora sob formas mais primitivas. Esta perspetiva histórica pode incrementar a nossa compreensão acerca da natureza transitória da existência humana na Terra e da resiliência das formas de vida que coexistiram conosco (algumas durante milhões de anos) e que podem continuar a sobreviver, mesmo que nós, humanos, desapareçamos. Provavelmente, se a espécie *Homo sapiens* se extinguisse hoje, as demais espécies do planeta talvez enfrentassem um período de adaptação, mas,

---

[6] Para mais informações sobre questões relacionadas com a regulamentação da IA, veja-se, por exemplo, Dumouchel 2023, 1023-1035.

em última instância, não sofreriam nenhuma ameaça existencial com a nossa ausência. Esta constatação deveria servir como ponto de inflexão ética, levando-nos a repensar o nosso papel no ecossistema planetário — não como centro, mas como parte interdependente de uma complexa rede de vida. Neste contexto, o desenvolvimento da IA torna-se particularmente relevante, uma vez que, à medida que a sua evolução continua a manifestar-se a alta velocidade, ela reflete a tendência humana para ultrapassar limites, para expandir o seu domínio, reencenando uma lógica histórica de controle, apropriação e instrumentalização da natureza e do homem nos mais diversos domínios da nossa sociedade. Poder-se-á dizer que a IA nasce, em grande medida, do desejo desmedido de se amplificar o poder humano, mas que está agora a ser convocada para conter os efeitos colaterais dessa mesma expansão. Surge, assim, um paradoxo: utilizamos uma tecnologia gerada por um modelo civilizacional predatório para tentar corrigir os próprios danos que esse modelo infligiu. De qualquer forma, as consequências do uso da IA refletirão, sem dúvida, a nossa atitude em relação a outras formas de vida na Terra e a nossa responsabilidade pelos ecossistemas que alteramos tão profundamente. É precisamente neste ponto que a proposta de Donna Haraway se revela particularmente interessante. Em *Staying with the Trouble: Making Kin in the Chthulucene*, Haraway desafia as narrativas dominantes do Antropoceno e propõe a ideia de «criar parentescos» (*make kin*) como alternativa à lógica antropocêntrica e tecnocrática que historicamente estruturou as relações entre humanos, outras espécies e o planeta. Para Haraway, «criar parentescos» significa assumir uma responsabilidade ampliada com formas de vida humanas e não humanas, promovendo alianças multi espécies, além da descendência genética e do vínculo utilitário. Transposta para o contexto atual de desenvolvimento acelerado da IA, a proposta de Haraway convida-nos a uma reconfiguração ética e política das tecnologias emergentes. A IA, se desvinculada de uma racionalidade extrativista e instrumental, pode ser reimaginada como uma ferramenta potencial para mitigar os danos ambientais que a espécie humana infligiu a outras formas de vida — especialmente num momento em que a crise climática e a perda de biodiversidade atingem proporções sem precedentes no nosso planeta. Em vez de continuar a expandir o domínio humano de forma hegemónica, a IA poderia ser convocada na reparação de ecossistemas, na preservação de espécies e na regeneração dos laços fragilizados entre humanidade e ambiente. Como destaca Haraway, «[i]t is not about finding a way out of the troubled times we inhabit, but about learning to stay with the trouble, to be truly present, to learn to live and die well with each other on a damaged planet».[7] Esta visão sugere um modelo de coabitação e interdependência que desloca o foco do progresso técnico para o cultivo de vínculos, de cuidados e de responsabilidade compartilhada. Nesse horizonte, o uso ético e criativo da IA poderia representar não apenas uma inovação tecnológica, mas um gesto ecopoético, ou seja, uma tentativa de reintegrar o humano numa teia de vida mais ampla, em que outras espécies — vegetais, animais e microbianas — fossem reconhecidas como agentes legítimos de um mundo comum.

O *Homo sapiens* foi programado pela evolução a concentrar-se no perigo. Durante milénios, a sobrevivência humana exigiu atenção imediata ao que ameaçava a vida de forma visível e palpável: o predador oculto, o veneno, o inimigo à espreita. A biologia moldou-nos para reagir com rapidez ao que grita, ao que salta aos olhos, ao que desperta o instinto de fuga ou de defesa — uma herança neurológica que privilegia o alerta diante do ruído, mas que, na verdade, pouco nos prepara para o silêncio. E neste contexto natural, o problema é que os perigos mais prementes da contemporaneidade não rugem nem sangram. São silenciosos. Codificam-se. Calculam-se. Programam-se. No espanto nutrido pelo algoritmo, a IA não é entendida como uma ameaça, apesar de embutida em aplicativos, sistemas de vigilância, bancos de dados, redes sociais e serviços automatizados. Confortável, eficaz e (quase) invisível não aciona os nossos alarmes evolutivos; não nos assusta como um incêndio, mas consome com igual fúria os alicerces do humano: a autonomia, a privacidade, a linguagem, a ética, a própria ideia de consciência. Por mais incrível que possa parecer, o *Homo sapiens*, ajustado para escapar aos lobos, abriga-se no silêncio do *software*, desvelando uma ironia profunda — tememos o que já dominamos e confiamos no que não compreendemos. Os algoritmos que filtram a realidade, sugerem ideias,

---

[7] Haraway 2016, 10.

influenciam comportamentos, ditam (in)justiça, promovem visibilidade, permanecendo, tranquilamente, fora do nosso radar intuitivo porque não mordem, mas amordaçam. O lobo já não está à nossa porta. Está em casa e foi convidado a entrar com a suavidade de um clique quase silencioso. Retomando a ideia de que vivemos num tempo onde a ideia de humanidade é posta em causa pela aceleração tecnológica, pela erosão da empatia e pela desmaterialização da experiência, somos confrontados com a urgência de reencontrar valores que reestruturem o ser humano antes que se dissolvam nos circuitos de dados e nas lógicas algorítmicas.

É neste contorno que a leitura de *Seis propostas para o próximo milênio*, de Italo Calvino (1923-1985), se torna uma fonte de atualidade e inspiração. Embora escritas no final dos anos 80, as seis qualidades que Calvino propôs como essenciais para a literatura (e, implicitamente, para a condição humana diante do futuro) — leveza, rapidez, exatidão, visibilidade, multiplicidade e consistência (não concluída) — podem hoje ser lidas como um mapa de resistência ética, estética e existencial frente à lógica desumanizadora da tecnologia. A leveza não se identifica como fuga da realidade, mas como uma forma de enfrentar a gravidade do mundo sem ser triturado por ela. A rapidez não se reconhece como aceleração cega, mas como pensamento ágil e crítico num mundo de sobrecarga informacional. A exatidão revela-se como um antídoto contra a linguagem vazia e as falsas narrativas que proliferam no ruído digital. A visibilidade enfatiza a capacidade de imaginar e dar forma ao invisível, ou seja, aquilo que as máquinas ainda não compreendem: a dor, o amor, o desejo. A multiplicidade é entendida como uma abertura a complexidades e contradições, contra o pensamento binário dos algoritmos e, por fim, a consistência que Calvino deixou por escrever. A bússola deixada por Calvino desenhada ao longo das suas propostas funciona não unicamente como uma série de princípios dirigidos unicamente para a literatura, mas também como um alinhamento de fundamentos para uma ética da presença, da imaginação e da responsabilidade no século da máquina. Neste sentido, a visão calviniana antecipa um modo de estar no mundo que exige atenção plena, responsabilidade criativa e uma imaginação capaz de restituir espessura ao que se torna cada vez mais raso, automatizado e descartável. Em pleno século da máquina, as propostas de Calvino oferecem um alicerce para repensarmos a nossa presença literária, ética e afetiva num mundo em transformação técnico-silenciosa.

Breves considerações finais: O espanto de (ainda) sermos humanos

> Dentro de nós há uma coisa que não tem
> nome, essa coisa é o que somos.
> Saramago 1995

Apontando para o mistério essencial da condição humana, as palavras de Saramago evocam o espanto de (ainda) sermos humanos num mundo que, cada vez mais, desafia as fronteiras do que isso significa. Entre algoritmos que simulam consciência, crises ambientais que expõem a nossa fragilidade e tecnologias que nos ultrapassam, permanece, no entanto, esse núcleo inominável sobre aquilo que somos ou que procuramos ser. Reconhecer esse ponto sem nome, esse centro móvel e inquieto, é talvez o primeiro passo para imaginar futuros menos mecânicos e mais habitáveis de tudo o que está fora do alcance das máquinas, ou seja, o afeto, o cuidado, o poético, a emoção. Perante este cenário, a pergunta «*Quo vadis*, humanidade?» não é apenas um eco retórico, mas um apelo urgente à reflexão. A inteligência artificial, quando transformada num fim em si mesma, revela-se como uma nova ideologia, um dogma que, sob a aparência da neutralidade e da eficiência, coloniza não apenas os espaços sociais e laborais, mas o próprio imaginário coletivo. Esta colonização não é visível à primeira vista — disfarça-se de progresso, veste-se de inovação e avança silenciosamente ocupando modos de ver, sentir, pensar, viver e criar. A lógica algorítmica torna-se o novo filtro da realidade, reorganizando a experiência humana segundo parâmetros de cálculo, controle e previsibilidade. O que antes era ferramenta passa a ditar os contornos do possível, do desejável e do aceitável. Neste sentido, não se trata apenas da necessidade de adaptação a uma nova era tecnológica, mas sim da necessidade urgente de resistir à tentação de reduzir o ser humano àquilo que pode ser automatizado para edificar o novo império colonial com a IA como capataz. No entanto, resistir não significa rejeitar o avanço; significa antes recusar a sua absolutização na defesa da pluralidade da experiência humana frente à homogeneização das respostas geradas por sistemas programados. Resistir é reafirmar que a inteligência, para ser plenamente humana, precisa incluir o erro, o silêncio, a hesitação, o gesto, a dúvida, o amor, o espanto — tudo o que escapa ao algoritmo. A problemática que se anuncia já não é o de parar a «máquina» porque não é possível, mas o de tentar (ainda e sempre) recuperar a nossa capacidade de perguntar porque e para quê a «máquina» existe. Reaprender a ser humanos, em tempos de inteligências artificiais que fazem (quase) tudo, é, talvez, o maior gesto de liberdade que podemos oferecer ao futuro. A verdadeira urgência talvez não seja acompanhar o avanço das máquinas, mas sim reaprender o que significa ser humano. Não se trata, portanto, de recusar os avanços tecnológicos, mas sim de perspetivarmos através de um outro ângulo a tecnologia que deve servir a vida humana e não substituí-la, como sugerem as palavras sempre atuais de Hannah Arendt (1906-75) em *A Condição Humana*:

> A ação, com todas as atividades que lhe são congêneres, é a única que vai diretamente dos homens para os homens, sem a mediação das coisas ou da matéria, e corresponde à condição humana da pluralidade, ao fato de que — e isto jamais poderia ser alterado — os homens, e não o homem, vivem sobre a Terra e habitam o mundo.[8]

Para Arendt, o que define a humanidade não é apenas o trabalho, mas sobretudo a ação, a palavra, a pluralidade, ou seja, aquilo que nos inscreve no mundo comum como seres políticos, éticos e relacionais. Como nos recorda Byung-Chul Han, vivemos sob o imperativo do desempenho e da positividade, em que a falha é inadmissível, isto é, a fenda exata onde a máquina se afirma e o humano se esvazia em gigabytes de Random Access Memory falsos e obrigatórios a transbordarem de positivismo. Como adverte José Saramago, «[p]enso que não cegámos, penso que estamos cegos. Cegos que veem. Cegos que, vendo, não veem»[9] e, por isso, talvez estejamos perante uma nova forma de cegueira: uma cegueira 'tecnificada' e voluntária que entrega à IA aquilo que tornava o ser humano habitação única para a capacidade de hesitar, errar, cuidar, narrar, viver, amar. Resta um apelo: que o amor, a linguagem, o pensamento crítico, a arte e o cuidado não sejam automatizados — que permaneçam como gestos humanos de resistência, memória e reinvenção. A velocidade com que a IA está a ser

---

[8] Arendt 2007, 15.
[9] Saramago 1995, 310.

integrada em contextos educacionais, médicos, judiciais ou artísticos levanta questões que não podem ser respondidas apenas com entusiasmo ou com medo. É necessário um novo pacto social, ético e poético, que reconheça a importância de manter a máquina como suporte e não como substituto da sensibilidade humana. Precisamos de reaprender a diferença entre processar e compreender, entre prever e imaginar. E é aqui que as humanidades recuperam o seu papel vital, uma vez que nos ensinam a escutar as entrelinhas, a habitar os silêncios e a resistir à lógica do desempenho, alertando-nos que qualquer dependência se transforma numa servidão. No caso da IA, uma servidão paradoxal, onde a promessa de liberdade, de tempo, de esforço, de escolha vem acompanhada de novas dependências, mais subtis, mais profundas e mais graves.

Neste céu sem estrelas, a questão fundamental deixa de ser apenas técnica ou científica: ela é poética e política. Poética, porque diz respeito à forma como imaginamos e significamos o mundo; política, porque envolve decisões coletivas sobre o que preservar, transformar ou abandonar. E nova pergunta surge: Que humanidade estamos dispostos a preservar? Talvez o destino ainda esteja nas nossas mãos — se houver coragem para hesitar, escutar e reaprender a analisar como sugere Harari que, habilmente, compara a ingestão de informação ao hábito básico de comermos:

> A nossa responsabilidade é limitada, mas temos alguma responsabilidade individual. E aqui diria que precisamos de uma dieta de informação. Tal como fazemos uma dieta alimentar. Sabemos que comer demais não faz bem, a *junk food* também não faz bem, deixa-nos doentes. O mesmo se passa com a informação. Demasiada informação, essencialmente se for má, deixa-nos doentes e adoece a sociedade. Precisamos de mais *tempo* para digerir a informação [...] porque a mente não tem tempo para digerir, pensar, refletir ou meditar. [...] Até precisamos de períodos de jejuns de informação.[10]

Vivemos num tempo em que o ser humano se adapta com inquietante naturalidade ao absurdo. Se o futuro vier a ser modelado por inteligências não humanas, que valores desejamos preservar e como traduzi-los para um novo alfabeto do sensível? Perante um tempo que narcotiza a perceção e acelera o esquecimento, permanece a pergunta: se já não nos espanta o

---

[10] Transcrição das palavras de Harari 2024.

facto de ainda sermos humanos, quem ou o quê poderá despertar-nos?

A inteligência, na sua essência mais profunda, permanece um dos maiores enigmas da existência humana. Não se trata apenas de um conjunto de capacidades cognitivas, mas de uma força rara, instável, irrepetível, um clarão que, até hoje, só foi concedida de forma plena apenas à espécie humana, como observa Barrat: «[a] force so unstable and mysterious, nature achieved it in full just once — intelligence».[11] Perante esta singularidade, o comprometimento pelo uso da IA e pelos desdobramentos éticos que dela decorrem recai, inevitavelmente, sobre nós. E aqui reside o maior risco apontado por Harari: «Is there anything more dangerous than dissatisfied and irresponsible gods who don't know what they want?».[12] Somos, talvez, esses deuses menores — criadores de uma inteligência que talvez já não nos pertença. Ao gerar sistemas capazes de aprender, decidir e agir, colocamo-nos perante uma nova pergunta que já não é apenas técnica: O que fazer com aquilo que criamos? Esta interrogação não admite respostas fáceis, porque ela toca no cerne da condição humana. Inventamos máquinas para prolongar os nossos desejos, mas talvez estejamos apenas a acelerar o nosso próprio esvaziamento. Resta saber se ainda somos capazes de imaginar outro futuro — um futuro que não se curve ao automatismo, mas reencontre no intervalo, na dúvida e na linguagem partilhada o gesto mais radical de liberdade.

Bibliografia

Amaral, Ana Luísa. 2022. *O Olhar Diagonal das Coisas*. Lisboa, Assírio & Alvim.

Barrat, James. 2013. *Our Final Invention. Artificial Intelligence and the End of the Human Era*. New York, Thomas Dunne Books.

Arendt, Hannah. [1958] 2007. *A Condição Humana*. Roberto Raposo. Rio de Janeiro, Forense Universitária.

Blanchot, Maurice. 1987. *O Espaço Literário*. Rio de Janeiro, Rocco.

Dumouchel, Paul. 2023. «AI and Regulations». *AI* 4.4: 1023-35. ‹https://www.mdpi.com/2673-2688/4/4/52›

---

[11] Barrat 2013, 5.
[12] Harari 2015, 416.

Foucault, Michel. [1975] 1987. *Vigiar e punir: nascimento da prisão*. Petrópolis, Vozes.

Haraway, Donna J. 2016. *Staying with the Trouble: Making Kin in the Chthulucene*. Durham and London, Duke University Press.

Harari, Noah Yuval. 2024. «Todas as Palavras. Entrevista com Yuval Noah Harari». *RTP*, 29 novembro 2024. ‹https://www.rtp.pt/play/p12723/e812741/todas-as-palavras›

Harari, Noah Yuval. 2015. *Sapiens. A Brief History of Humankind*. New York, Harper Perennial.

Pattanayak, Shibabrata. 2021. «Homo Sapiens: Is the Species Making Itself Extinct?». *Academia Letters* 1: 1-6. ‹https://doi.org/10.20935/AL1617›

Saramago, José. 1995. *Ensaio sobre a cegueira*. Lisboa, Caminho.

Taylor, Charles. 2016. *The Language Animal: the Full Shape of the Human Linguistic Capacity*. Cambridge, The Belknap Press of Harvard University Press.

AI in the Humanities:
Perception, Persuasion, Pedagogy

Martha Brenckle & Patricia Farless

We often make self-deprecating jokes about our own learning, claiming that everything we know about Classical music was learned from watching Warner Bros. Cartoons. Only slightly true, we do think, however, that most people have come to know about artificial intelligence (AI) through films, television series, and science fiction texts. Popular culture, science and speculative fiction have bombarded us with stories and images where intelligent, thinking machines develop questionable ethical behaviors and become almost impossible to control. In the classic film *2001 A Space Odyssey*, the HAL 9000 supercomputer takes over the spaceship, killing all but one of the astronauts because it did not like being told it had made an error. HAL 9000 has a pleasing human voice, conducts conversations and makes 'small talk'. Even HAL's death takes on human pathos as it is forced back to its early programming when Dave Bowman turns off the computer as HAL begs Bowman not to disable it. HAL 'dies' singing «Daisy, Daisy, give me your answer do».

The *Terminator* movie franchise takes the fear of machines to the level of world annihilation. Using apocalyptic tropes, the future landscape is imagined as a burned-over wasteland, where the few humans left struggle to survive, as they war with Skynet, a hostile artificial intelligence. To stop humans from changing their future, a cybernetic assassin is sent back in time from 2029 to 1984 to assassinate Sarah Connor, whose unborn son will one day save humankind from extinction by presumably not inventing machines that can think for themselves.

The film *I, Robot* which was released twenty years after the first *Terminator* movie is loosely based on Isaac Asimov's short story collection of the same name which has a subtitle *Man-Like Machines Rule the World*. Included in the film is Asimov's

Three Laws of Robotics which act as an organizing principle and unifying theme of the film:

1.   A robot may not injure a human being or, through inaction, allow a human being to come to harm.
2.   A robot must obey the orders given it by human beings except where such orders would conflict with the First Law.
3.   A robot must protect its own existence as long as such protection does not conflict with the First or Second Law.[1]

Set in Chicago in 2035, highly intelligent robots fill public service positions throughout the world, operating under the Three Laws of Robotics to keep humans safe. Detective Del Spooner investigates the alleged suicide of U.S. Robotics founder Alfred Lanning and believes that a human-like robot called Sonny murdered him. Spooner believes that the Robot Laws can be misinterpreted by AI and that robots are dangerous.[2] Even if you haven't seen the film, you can easily imagine what happens next.

The problem we are faced with as humanists is not that a machine is capable of thinking like a human being, as we know that computers 'think' by making associations and retrieving data that we ask for, but it cannot think critically about the data it collects. Nor can it know the questions to ask. The problem of believing computers have human attributes is a lack of knowledge situated in human emotion and belief.

Since chess playing computers were designed, many people decided that not only do computers think like humans but that they are better thinkers than humans and will someday replace us. The famous chess match in 1997 when IBM's AI, Deep Blue beat Gary Kasparov, arguably the chess world's best and most famous Grand Master, cemented the belief that computers are smarter than people. In a six-game series held in New York City, Deep Blue won the match in just 19 moves. A photo taken after the match shows a very stone-faced Kasparov barely acknowledging the handshake he received from Dr. C. J. Tan, head of the IBM Deep Blue computer team. C. J. Tan is triumphantly smiling. There is no corresponding photograph to show how Deep Blue was feeling.

It is relatively easy to make people afraid of AI; it is much harder to make them unafraid. For the most part, we would

---

[1] Asimov 2008.
[2] Proyas 2004.

say that those of us teaching in the Humanities disciplines are skeptical of AI; even the few pedagogical practices about using AI in the classroom have not been able to displace the worries about disciplines that depend on textual evidence, use writing to augment students' critical thinking processes, and to evaluate and assess student work. We spend hours trying to find out what AI cannot do, like write an argument about disparate pieces of evidence, or understand and take a position on an issue. AI makes factual mistakes, creates references that do not exist, takes on the biases of its programmers and the texts that have been fed into its memory. Currently, AI is not conscious of itself or nor does it have the violation to become a destructive, self-aware entity like *Terminator's* Skynet. What AI can do very well is operate as a powerful data sorting mechanism that works at impressive speeds. It can recognize patterns in large data sets, conduct complicated mathematical calculations, and analyze information faster than its human counterparts. AI has many possible applications in engineering and health, architecture, city planning, agriculture, and construction industries. Students studying in those areas will learn to use AI in their professional lives.

So, what is the role of those of us who teach and work in the Humanities, who know the significance of art, music, philosophy, rhetoric, languages, and history in understanding humans individually and in communities? To us, human expression, creativity, ethics, and communication are significant factors in developing empathy and critical thinking. How does AI fit into our academic work? We do have a few choices; we can read books and articles about how to use AI in our teaching, or we can dig ourselves deeper into our disciplinary caves and pretend AI isn't happening. This does, however, make us remember how when teaching online was first introduced, the image of *The Great Wave Off Kanagawa*, a wood block print created by Japanese artist Hokusai in 1831, was reproduced and put in front of teachers as a warning: you have no control, so ride the wave or drown.

Hiding or drowning is not an option most teachers would take as we know that AI is now part of the landscape of our students' lives, and it is our job to help them get ready for their futures as world citizens. Rather than just ride the wave, we can do what we did with online teaching, Wikipedia, citation generators, and other technologies: we can think about it critically, learn its history, how it works, what it can do and not do, and how our students might make the best use of newer technologies. Let's look at AI with our disciplinary lenses and ask ourselves what do we need to know about AI to feel more in control than the image of the tsunami wave would lead us to believe?

First, we need to get past the conversations about plagiarism. Students have always cheated writing papers, and the ability to cut and paste has made plagiarism even easier. Yes, they can ask the many AI programs to write their papers but how is that different from buying papers or having a friend or relative write it? If we focus on plagiarism, it becomes a long, downward spiral with no end in sight. This conversation also feeds into the thinking of those in the academy who continually remind us that in the future, no one will need to learn to write because AI will do all the writing anyway. We have even been told that our disciplines will no longer be necessary.

Second, we need to reject the now cliched «Digital Natives and Digital Immigrants» concept. Yes, we had rotary phones when we were young, and our students have grown up with digital technologies. But we have found that students know how to digitally do what is interesting to them, but not necessarily what they need to know to be professionally successful. Our students know how to use Canva to make posters and memes, but still need to learn theories of visual rhetoric, as well as art and design principles, so they can communicate effectively with images. Students need to be able to develop worthwhile research questions and topics; asking the pertinent questions is the most difficult part of the research process. When students are researching online, they need to be able to evaluate the sources they find and then narrow down their searches. University students going into well-paying careers need to know more than how to use AI; they need a firm understanding of how AI thinks, how algorithms work, and how to ask questions to receive the information they need.

How do we prepare students for an AI driven future so they can navigate the complex interplay of human and AI interactions across the various roles they will play as individuals, citizens, and professionals, while working with the knowledge, skills, and human values that are important to Humanities disciplines? Fawzi BenMessaoud, the AI Program Director at Indiana University-Indianapolis, has developed a list of competencies and skills students will need in a world where AI is pervasive, imbedded in landscape of our daily lives, and working with AI is the digital norm. Dr. BenMessaoud writes, «I propose a tripartite model of competencies and skills integral to thriving in an AI-driven world: intelligent design skills, intelligent human skills, and intelligent data skills».[3] The AI skills students learn should not be additive but should be imbricated in communication, cultural awareness, creative expression. In this way, as BenMoussad writes, «Education should not merely react to technological shifts but actively shape them».[4] According to BenMoussad, these crucial skills and competencies include:

Intelligent Design Skills: The Creative Synthesis

The first sector of the model articulates intelligent design skills — logical mind mapping, which represent the creative *harmony between human aspiration and technological feasibility*. They empower individuals to craft solutions where form meets function and functionality, *embedding user-centric principles* into the heart of AI solutions.

Instrumental and Critical Digital Competence

A profound literacy in digital tools and a discerning approach to digital content and ability to effectively use technology are indispensable. This competence underlines the necessity for individuals to discern and manipulate digital resources innovatively and responsibly within AI paradigms, logical *mind mapping*, an understanding *of systems* and how they interact, *creative design*, and *empathy in aligning AI systems with human needs and experiences*.

Intelligent Human Skills: The Ethical Imperative

This second facet of the model describes intelligent human skills, which are the quintessence of the human within the digital realm. *Reflective, critical thinking and adaptability* are not luxuries but necessities, equipping individuals to partner with AI, ensuring technologies are wielded with *wisdom and sensitivity*. Intelligent human skills focus on personal competencies that are uniquely human and may be beyond AI's capability to replicate. These include *assessment and decision-making competence, ethical competence, reflection, and learning skills*.

---

[3] BenMessaoud 2024.
[4] BenMessaoud 2024.

Intelligent Data Skills: The Analytical Cornerstone

> The third segment, intelligent data skills, covers new litera-
> cies and highlights the importance of data-centric compe-
> tencies in leveraging AI. In an information drenched land-
> scape, the *ability to interpret, analyze, and apply data* trans-
> cends traditional academic boundaries. These skills enable
> individuals to transform data into decisions, driving inno-
> vation and uncovering insights within vast information
> streams. They are now foundational in this data-driven
> landscape and include the ability to *verify sources, translate
> information into coherent narratives, turning data into im-
> pactful visuals in order to effectively communicate complex
> information driven solutions, archive and curate infor-
> mation to preserving its relevance and ensuring its utility
> and quality over time for ongoing and future AI initiatives.*[5]
> (emphasis added)

The emphasis within BenMoussad's eloquent list of necessary skills is all ours. The italics are there to remind Humanities professionals that we already value and teach these skills
in our disciplines. What students will learn in our classrooms
is how to best use and represent AI solutions and merge human aspiration and technology. That may start with using AI
to brainstorm or connect ideas and concepts but will incorporate human needs and values. But as instructors, we have a
responsibility to have an overview of how AI in fact works. We
asked ourselves a few questions and then did what we do best;
we conducted research to find the answers.

How Do Computers Think?

The earliest substantial work in the field of artificial intelligence was done in the early to mid-20th century by the British logician and computer pioneer Alan Mathison Turing. In
1935 Turing described an abstract computing machine consisting of a limitless memory and a scanner that moves back and
forth through the memory, symbol by symbol, reading what it
finds and writing further symbols. The actions of the scanner
are dictated by a program of instructions that also is stored in
the memory in the form of symbols, and implicit in it is the
possibility of the machine operating on, and so modifying or
improving, its own program.

---

[5] BenMessaoud 2024.

In 1950, Allan Turing was asked to consider the question of artificial intelligence (AI) by working to answer the question «Can machines think?»

> I propose to consider the question, «Can machines think?» This should begin with definitions of the meaning of the terms «machine» and «think». The definitions might be framed so as to reflect so far as possible the normal use of the words, but this attitude is dangerous, If the meaning of the words «machine» and «think» are to be found by examining how they are commonly used it is difficult to escape the conclusion that the meaning and the answer to the question, «Can machines think?» is to be sought in a statistical survey such as a Gallup poll. But this is absurd. Instead of attempting such a definition I shall replace the question by another, which is closely related to it and is expressed in relatively unambiguous words.[6]

Turing's conclusion after rejecting at least seven arguments and solutions from the perspectives of disciplinary thinking was that sufficient complexity in building machines would allow them to mimic human thinking. A quotation often attributed to Turing (but found nowhere in his published work) surmised that a computer would deserve to be called intelligent if it could deceive a human into believing that it was human.

The key words are «mimic» and «deceive». As most people who work with GenAI in this decade do not attribute an ability «to deceive or lie» (a choice based on moral or religious evidence) to AI but rather call its miscalculations and errors «hallucinating»; rather, we would like to focus on Turing's possible use of the word «mimic». Mimic implies a performance, a resemblance to someone or something like when a butterfly «mimics» a leaf. A butterfly may look like a leaf in its ability to hide from predators, but it will never be a leaf. AI may imitate human thinking, but it isn't actually processing or thinking in the same way humans do.

AI 'thinks' by recognizing patterns and making associations. The ability to solve problems and come to logical conclusions is one method of critical thinking assigned to human thinking. The process of critical thinking goes beyond accepting information at face value and understanding how to apply

---

[6] Turing 1950, 433.

it effectively and ethically in various contexts. It's also considered a form of emotional intelligence, as it requires insight into one's own biases and assumptions. The psychologist H.E. Gardner defines intelligence as the «biopsychological potential to process information that can be activated in a cultural setting to solve problems or create products that are of value in a culture».[7]

Currently, training a Large Language Model (LLM) to reason like a human is one of the biggest challenges in AI research. By applying Chain-of-Thought prompting, Self- Consistency, and Retrieval-Augmented Reasoning, researchers may help AI move from blindly predicting text to logically solving problems. As Turing writes: «I believe that at the end of the century the use of words and general educated opinion will have altered so much that one will be able to speak of machines thinking without expecting to be contradicted».[8]

Although AI can perform or mimic human-like behaviors, it doesn't think or understand physical reality like humans. It is not conscious of itself. Artificial general intelligence (AGI) is the point at which AI acquires human-like generalized cognitive capabilities, but lacks true understanding, embodied experiences, and emotions, making it fundamentally different from human intelligence. Creating human-like AI is about more than mimicking human behavior — technology must also be able to process information, or 'think', like humans too if it is to be fully relied upon.

In mimicking how humans learn, AI learns through explanation, simulation, analogy, and reasoning without external inputs. This on-demand learning, beneficial for adapting knowledge to new contexts, illustrates similarities and pivotal differences between natural and artificial cognition, offering a unique lens to study human thought processes and AI's potential and limitations. AI can only think with human input.

According to Tania Lombrozo, a professor of psychology and co-director of the Natural and Artificial Minds Initiative at Princeton University, «AI has gotten to the point where it's so sophisticated in some ways, but limited in others, that we have this opportunity to study the similarities and differences be-

---

[7] Gardner 2000, 28.
[8] Turing 1950, 460.

tween human and artificial intelligence. We can learn important things about human cognition through AI and improve AI by comparing it to natural minds. It's a pivotal moment where we're in this new position to ask these interesting, comparative questions».[9]

ChatGPT only became useful with the addition of what's called «reinforcement learning from human feedback» (RLHF):

1. The model produces outputs
2. Humans rate those outputs for helpfulness
3. The model is adjusted in a way expected to get a higher rating

A model that's under RLHF hasn't been trained to only predict next words, it's been trained to produce whatever output is most helpful to human raters. «Think of the initial large language model (LLM) as containing a foundation of knowledge and concepts. Reinforcement learning is what enables that structure to be turned to a specific end»,[10] writes Benjamin Todd in a blog post «Teaching AI to reason».

Now AI companies are using reinforcement learning models to teach AI to reason step-by-step:

1. Show the model a problem like a math puzzle
2. Ask it to produce a chain of reasoning to solve the problem («chain of thought»)
3. If the answer is correct, adjust the model to be more like that («reinforcement»)
4. Repeat thousands of times

This process appears to be working in mathematics and legal reasoning. And to a large extent, the process of teaching AI to reason sounds a lot like what happens in our classrooms, especially the repetition thousands of times.

How do algorithms work?

Even if we don't know how to write computer codes, we do know that AI uses algorithms. Algorithms are fundamental to computer science and are used in various fields such as mathematics, data science, and artificial intelligence. An algorithm is a set of well-defined instructions or rules designed to solve a specific problem or perform a computation. It is a step-by-

---

[9] Lombrozo 2024, 1015.
[10] Todd 2024.

step procedure that takes an input, processes it, and produces an output. To work, algorithms must be clear and unambiguous, have well-defined, specific inputs and outputs, have finiteness so it terminates after a set number of steps, the steps be executed with the available resources, and the algorithm can be implemented in any program language. Algorithms are a set of instructions much like giving a friend directions to your house or when you use your phone to find a restaurant you read about.

Like any language, algorithms can be biased. When we process information and make judgments, we are inevitably influenced by our experiences and our preferences. As a result, people may build these biases into AI systems through the selection of data or how the data is weighted. For example, cognitive bias could lead to favoring datasets gathered from Americans rather than sampling from a range of populations around the globe. Algorithmic bias occurs when systematic errors in machine learning algorithms produce unfair or discriminatory outcomes. It often reflects or reinforces existing socioeconomic, racial and gender biases. Algorithmic bias can be unfair or even dangerous when found in health care, criminal justice, or human resources. Teaching students how to evaluate sources and outcomes for possible bias will give them useful skills in using AI in their chosen professions.

What is Prompt Engineering?

When teaching research skills, we talk to students about asking the right questions in a way that will shape their work. How questions are asked is crucial for the efficient use of AI, so the models generate high-quality, relevant, and accurate responses. Input prompts must be crafted and refined to optimize the output of most AI models. A well-crafted prompt can significantly influence the AI's output, making it more coherent and meaningful. Prompt engineering involves understanding the nuances of language, context, and the specific requirements of the task at hand.

The primary benefit of prompt engineering is the ability to achieve optimized outputs with minimal post-generation effort. By crafting precise prompts, prompt engineers ensure that AI-generated output aligns with the desired goals and criteria, reducing the need for extensive post-processing. A successful prompt engineer must have strong communication skills, knowledge of programming languages, and an understanding of machine learning to predict user behavior and craft more useful prompts.

How will AI become part of the learning process?

According to Bowen and Watson, we should immediately flood our curriculum and classrooms with AI topics — ethics, personal information privacy, civics, environmental and energy costs, labor, discrimination, bias, economics, justice, and education. In our classrooms we perform a balancing act between teaching the content knowledge of our disciplines and teaching the processes that make knowledge we value. Now that content is readily available on the internet, we need to shift our focus from the «what» to the «how and why». «Our new future is teaching students how to think with AI».[11] Fortunately, unlike HAL in *2001 A Space Odyssey*, AI programs won't refuse to answer prompts or directions. And its feelings don't get hurt.

Conclusion

We asked Co-pilot to write a 50-word conclusion to this paper and in less than a minute, it sent us this:

> AI's integration into the humanities challenges traditional pedagogical practices, urging educators to critically engage with technology. By understanding AI's capabilities and limitations, humanities professionals can harness its potential to enhance critical thinking, creativity, and ethical reasoning, preparing students for a future where AI is ubiquitous.[12]

It's passable, but we are not sure if we were «challenging traditional pedagogical practices». Many of our practices will remain the same; they will be enhanced by AI rather than being significantly upended. We hoped we showed that most of our uneasiness comes from having to become comfortable with another technology — no one, especially instructors, enjoys being a beginner. We hope we managed to show how the Humanities is always already a significant facet of AI in areas of teaching research methods and writing processes. AI will become faster, more efficient and able to solve problems with

---

[11] Bowen and Watson 2024, 238.
[12] Response generated by Co-pilot.

logical reasoning but studying the Humanities will help any new technology to include the human element — ethical decision making, embodied experiences, empathy, and creative and emotional intelligence.

Bibliography

Asimov, Isaac. 2008. *I, Robot*. New York, Del Ray Publishers.

BenMessaoud, Fawzi. 2024. «Must-Have Competencies and Skills in Our New AI World: A Synthesis for Educational Reform». *EDUCAUSE Review*. 17 September 2024. ‹https://er.educause.edu/articles/2024/9/must-have-competencies-and-skills-in-our-new-ai-world-a-synthesis-for-educational-reform›

Bowen, José Antonio and C. Edward Watson. 2024. *Teaching with AI: A Practical Guide to a New Era of Human Learning*. Baltimore, Johns Hopkins University Press.

Cameron, James. 1984. *The Terminator*. Los Angeles, Orion Pictures.

Gardner, Howard. 2000. *Intelligence Reframed: Multiple Intelligences for the 21st Century*. London, Hachette Press UK.

Kubrick, Stanley. 1968. *2001: A Space Odyssey*. Burbank, Warner Brothers Pictures.

Lombrozo, Tania. 2024. «Learning by Thinking in Natural and Artificial Minds». *Trends in Cognitive Sciences* 28.11: 1011-1022.

Proyas, Alex. 2004. *I, Robot*. Los Angeles, 20th Century Fox.

Todd, Benjamin. 2024. «Teaching AI to Reason». *Benjamin Todd*. 11 January 2025. ‹https://benjamintodd.substack.com/p/teaching-ai-to-reason-this-years›

Turing, Alan M. 1950. «Computing Machinery and Intelligence». *Mind* 49: 443-460.

Intermediate Level Writing in Portuguese and the Roles of Creativity and GenAI*

Flávia Azeredo Cerqueira & Eduardo Viana da Silva

Introduction

This chapter looks at the potential and limitations of creativity and the use of generative artificial intelligence (GenAI) in the writing production of intermediate-level students of Portuguese as an additional language (PAL). Under the umbrella of artificial intelligence (AI), generative AI is a tool that creates images, audio or video based on existing data and learned patterns. This chapter explores the roles of creativity and GenAI in the writing process of 34 university students of Portuguese at two large institutions of higher education in the United States. It seeks to explore how digital technologies, such as ChatGPT, may contribute to, or hinder, students' writing and their ability to create with language.

Creativity lies at the heart of the language learning process. A distinguished trait of intermediate-level speakers is their ability to create with language. The ACTFL Language Connects 2024 Writing Proficiency Guidelines describes intermediate-level writers as follows:

> Writers at the Intermediate level have the ability to meet practical writing needs, such as simple messages and letters, requests for information, and notes. In addition, they can ask and respond to simple questions in writing. These writers can create with the language and communicate simple facts and ideas in a series of loosely connected sentences on topics of personal interest and social needs. They write primarily in present time. At this level, writers use basic vocabulary and structures to express meaning.[1]

* The authors thank the editors, Susana Antunes and Sandra Sousa, for their invitation to contribute to this important volume. We are also grateful to our students for participating in the case study. Obrigada!/ Obrigado!

[1] ACTFL 2024.

The writing skills of intermediate level learners are marked by their ability to combine and recombine language structures and vocabulary in the additional language. By creating with language, learners at the intermediate level differentiate themselves from beginning language students. The written production at the intermediate level, as pointed out in the excerpt above, is primarily done with loosely connected sentences and with more accuracy in the present time. When dealing with narratives in the past and detailed descriptions, the quality and quantity of language generally decrease significantly. Nonetheless, intermediate speakers have functional abilities. They are able to produce language at the sentence level, with discrete sentences and/or strings of sentences. At times, their writing is characterized by skeletal paragraphs and/or a combination of short paragraphs for stronger performances at the intermediate level.

For the purposes of this chapter, we adopted the proficiency descriptors from ACTFL Language Connects, a U.S. organization that provides guidelines on language proficiency for speaking, listening, reading, and writing. ACTFL is well known in the U.S. for its Oral Proficiency Interview (OPI), which determines the language level of learners through real-time interviews. In the realm of writing, ACTFL has the Writing Proficiency Test (WPT), which analyzes the writing ability of learners based on the completion of writing tasks, considering the content areas and contexts, accuracy of the writing, the length, and organization of texts produced.

Given the increasing number of students and faculty using GenAI, this chapter focuses on the impact of ChatGPT as a tool for the revision of writing assignments and for suggestions on expansions of such assignments. We are interested in understanding how ChatGPT may help and/or hinder the writing production of PAL students. We followed a set of principles in the design of the case study. The first principle is that the initial piece of writing should be produced by students using pen and paper only, without the assistance of GenAI. The second principle is that GenAI emphasizes a more individualized learning process, instead of a more social learning process with the collaboration of other students (as in peer reviews for example). Finally, our third principle is that both GenAI and peer review processes have potentials and limitations. Our research question focuses on the use of GenAI, and specifically ChatGPT:

How may ChatGPT contribute and/or hinder the students' ability to review their writing and to create with language? The next section presents a brief background on the roles of creativity and GenAI in writing.

Language Creativity and GenAI

In his book edition *The Routledge handbook of language and creativity*, Jones identifies three elements involved in the phenomenon of linguistic creativity, namely: semiotic resources, the cognitive processes, and the social processes, as represented in the figure below.

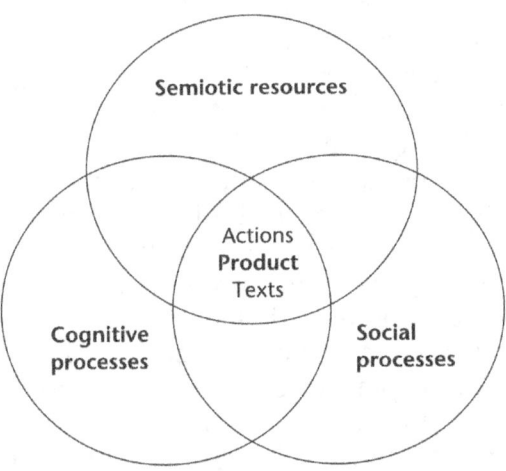

Figure 1. Conceptual map for language creativity (Jones 2015).

The texts produced by language students are therefore the result of the semiotic resources available to them as the writing process by pen and paper, for example, and/or the use of electronic dictionaries, rubrics, and technological tools, such as ChatGPT or Google translation. The cognitive processes involve the ability of writers to recall information, make decisions on possible feedback (what to take and what not to take), and their ability to reason and perceive what is appropriate or not in their writing. Finally, the social processes for writers are connected to social contexts and target audiences. By contextualizing writing tasks and having a clear audience, writers tend to adapt their writing and make changes accordingly. We also argue that in the case of writing processes that include peer review, there is a social component to it, as writers take it into account the readers participating in the peer review. If there is a conversation between the writers and the peer reviewers, this is another social component of the writing process. The same does not occur necessarily in this manner with GenAI tools, such as ChatGPT, which tends to be a more individualized learning process.

According to Jones' descriptions of language creativity, language makes available some affordances and is also guided by constraints. The affordances in language creativity refer to «the ability to combine a limited number of elements into different patterns and to mix it with other modes to increase its meaning potential».[2] Applied to the study presented in this chapter, affordances are related to the language knowledge in Portuguese that students already acquired and their ability to mix and match vocabulary and language structures in order to produce their written texts. The constraints would be the limits that the Portuguese language imposes in its lexical and syntax; in other words, the limitations imposed by grammatical rules of Portuguese and lack of vocabulary from the students, for example. Research on creativity and second language (L2) learning has demonstrated a relationship between access to an extensive mental lexicon and the creative use of language.[3] According to Kormos, «creativity plays a significant role in the content planning, idea generation, organization, and lexical selection processes of L2 writing».[4]

At the same time that having access to more vocabulary facilitates the creative process in writing, having more fluency in the L2 also facilitates the automatization in the language production. Therefore, a writing task at the lower levels of proficiency is cognitively more demanding as the L2 writer deals with the selection and retrieval of vocabulary, the syntactic and morphological encoding and spelling procedures,[5] among other factors that slow down the writing process and impose challenges in planning and revising texts. Chenoweth and Hayes explain that L2 writers might need to simplify their writing goals because of the overload of attentional resources.

---

[2] Jones 2015, 7.
[3] Fernández-Fontecha 2021; Fernández-Fontecha 2022.
[4] Kormos 2024, 2.
[5] Kormos 2024; Stevenson 2007.

The writing production of intermediate level students is analyzed in the case study following the guidelines of the Writing Proficiency Test (WPT) from ACTF Language Connects 2024, as presented in the table below.

Proficiency level	Task and Functions	Context/ Content	Text Type	Accuracy
Intermediate	Can meet practical writing needs, i.e. notes, simple messages, and requests for information. Can ask and respond to straightforward questions.	Routine informal setting and limited tasks involving exchange of simple information. Predictable, familiar topics related to self and daily routines and activities.	Writes a loosely connected text made up of a connection of primarily discreet sentences that may or may not be presented in the semblance of a paragraph.	Express meaning through vocabulary and basic structures that is comprehensible to those accustomed to the writing of non-natives.

Table 1. Writing Proficiency Test Assessment Criteria (2024)

The table above describes the expectations of intermediate level writing skills in an additional language. At the intermediate level class, topics presented in writing correspond to those learned previously in class or other learning settings, being predictable and familiar to the language learner. The vocabulary and language structures are also related to what students have learned before. Typically, intermediate language learners of Portuguese are expected to have more control of tasks in the present tense, while struggling to produce past narrations, either because of a lack of command in the language and/or adequate vocabulary, and lack of accuracy of language structures, verb conjugations, among other aspects. Stronger performances at the intermediate level may exhibit high accuracy in past narratives; however, learners are generally unable to maintain this level of proficiency consistently across tasks.

In using GenAI as a tool in language writing, we must consider the pros and cons of it. What are some of the benefits of using GenAI? What are some of its affordances and constraints? To what extent should language instructors and language learners embrace and/or refuse the use of GenAI?

McKee (2025) proposes the intersection between innovation, adaptation, and resistance when dealing with GenAI tools in writing. In other words, the stakeholders in language learning must decide how to find a balance between innovating and adapting the teaching of writing skills in relation to GenAI or resisting its use at times. The premise of the current study is that GenAI, in particular the generative AI tool ChatGPT, could be successfully used as part of the writing process of PAL students. Nonetheless, there are reservations in how this GenAI tool could be applied to the writing process.

Researchers have argued against the use of GenAI in academic assignments.[6] Among the arguments against the use of GenAI in writing, Sano Franchini, MacIntyre, and Fernandes established ten premises, including a stand against linguistic homogenization fomented by GenAI, the fact that GenAI is never ideologically neutral, and the range of labor issues that come with the adoption of GenAI. Their arguments are strong and valid. Nonetheless, the case study presented in this chapter does not defend the use of GenAI for the production of texts from its initial stages. Instead, we propose the use of GenAI for providing corrections in texts and suggestions on how to expand them creatively. We argue that GenAI has a place in the language classroom if used consciously and cautiously. Although ChatGPT was the GenAI tool adopted in the case study presented in this chapter, other GenAI tools could also be used instead, such as Gemini and Copilot.

GenAI works with different language models in order to produce and/or revise written texts, namely: Natural Language Processing (NLP), Natural Language Understanding (NLU), and Natural Language Generation (NLG). McKee explains that ChatGPT has 1.3 to 1.8 trillion parameters (GPT-4), which function as artificial neurons building connections. Each connection is a parameter allowing for 'understanding' data in context. Despite its benefits in providing immediate feedback and formative assessment for writers, McKee points out the many shortcomings of ChatGPT, such as factual errors, distortions, violations of copyrights, bias, the sharing of data, environmental impact, homogenization of language and thinking, and the risk of loss of agency in human thinking and learning.

---

[6] Bender 2024; Bender 2021.

Far from ignoring the risks of GenAI in the educational process of PAL, we look for possible applications of it, which may benefit students. At the same time, we are conscious of the risks imposed by GenAI and these should not be underestimated. We follow the educational framework for GenAI proposed by Bedington, where GenAI is used as an assistant in the writing process for idea developing, editing, and evaluating pieces of writing produced by language learners. We purposely do not follow the applications of GenAI for composing, which is proposed by Bedington as a writing assistant. We believe that it would be too much of a fine line for language students at the intermediate level (or perhaps at any level) to adopt texts produced by ChatGPT. The risk is that instead of using it as a tool, students could adopt ChatGPT as the main source for the production of texts. We believe that would be a mistake. Nevertheless, Bedington's metaphor of the GenAI as a bicycle seems to fit the premises that we follow in the case study presented in this chapter:

> I was initially worried that AI would be like a car, removing all exercise for me as a writer and thus not letting me develop and exert myself. But now I see AI as a bicycle. Just as a bicycle can take me farther than if I walked but it still is *my exercise* that is shaping the experience, so now I see how AI can improve my writing without replacing the mental exercise of the writing process.[7] (emphasis added)

As language instructors and students engage with writing tasks nowadays, there is an increasing need to foster digital literacies. Jamielson defends the position of trusting students and to instruct them on AI processes: «A pedagogical response calls on us to trust students; to teach them the work of writing and include AI in the process».[8] According to Selber, digital literacy is divided into: (1) Functional literacy (computers as tools and users of technology), (2) Critical literacy (computers as cultural artifacts and users as questioners of technology), and (3) Rhetorical literacy (computers as networks and user as producers of technology). In this book chapter, we bring the digital literacy perspective to the forefront while incorporating the use of GenAI in the writing process of PAL students. The next sessions will explain the methodology applied to a

---

[7] Bedington 2024, 8.
[8] Jamielson 2022, 156.

case study with 34 university students taking PAL in the United States, while analyzing the efficacy of using GenAI in the writing process and its role in writing creatively as well.

Methodology

Case Study

The case study presented in this chapter investigates the impact of GenAI (ChatGPT) on the writing production of Portuguese, focusing on intermediate-level learners of PAL. The case study was designed using a structured experimental approach, consisting of a pre-test, an intervention phase, a post-test, and a survey. The primary objective of the study is to evaluate how the use of GenAI affects the writing skills of intermediate PAL learners. Key areas of focus include vocabulary usage, grammar accuracy, coherence, fluency, and overall writing quality. An additional, more exploratory goal is to assess the potential impact of GenAI on the learners' creativity during the writing process. By examining these aspects, the study aims to provide insights into the practical applications of GenAI tools in PAL learning, particularly for writing tasks. It seeks to understand not only the technical improvements in language use but also how these tools influence creative expression and the development of individual writing styles.

Participants

The case study included 34 participants, all of whom were enrolled in intermediate-level Portuguese language courses at two large universities located in the U.S. East Coast ($n=20$) and West Coast ($n=14$). The participants ranged in age from 19 to 27 years.

Among the 34 participants, the linguistic backgrounds varied. Some participants were monolingual speakers of English ($n=8$). Although these individuals had been exposed to Spanish throughout their academic lives, they did not possess functional proficiency in Spanish. One participant was bilingual in Chinese and English and two were heritage speakers of Brazilian Portuguese, while the remaining 23 participants were bilingual in Spanish and English. Participation in this study was voluntary, and all participants were selected based on their language background, as well as their enrollment in an intermediate Portuguese language course at their respective universities. There was no financial compensation offered for participation in the research.

Pre-test (Baseline Assessment. First writing draft by pen and paper)

At the outset of the study, both groups of participants (East Coast and West Coast, n=34) were asked to complete a writing task as part of the baseline assessment. The participants were instructed to produce their responses with pen and paper, under controlled conditions, without the aid of GenAI tools or any other external resources. Specifically, they were prohibited from using dictionaries, textbooks, or class notes. This restriction ensured that their initial writing abilities were assessed independently of any external support. This writing task was completed in class and participants (n=34) had 30 minutes to produce their texts. The prompt for the writing task was titled *Seu Fim de Semana Prolongado (Your Long Weekend)*. Students were asked to write a blog entry targeted at a Portuguese-speaking audience, reflecting on their university life and recent experiences. The specific prompt read:

> Write about your last weekend. What did you do during the holiday weekend (Dia do Presidente)? Did you go somewhere? Did you stay home most of the time? Did you talk to friends or family? Write at least 8 complete sentences and organize your composition into cohesive paragraphs.

This pre-test was designed to assess the participants' baseline writing skills in several key areas, including vocabulary, grammar, coherence, and overall organization. The class instructors reviewed the essays using a symbol system, where inaccuracies were circled in the text, but not corrected (e.g. VF for Verb Form, where the verb «foi» is circled in the sentence «Eu foi para casa»). The results from this initial task provided a point of comparison for evaluating the impact of the GenAI-assisted and Peer Reviewed interventions later in the study.

Intervention Phase. Review for linguistic accuracy (n=34)

During the intervention phase, students were tasked with reviewing and revising their initial writing in class for 25 minutes. To ensure a balanced approach, participants were randomly divided into two groups, each following a distinct revision method:

Group A (Peer Review Group) (n=16). The first half of the participants (Group A) engaged in a peer review process. In this control group, students worked in pairs, exchanging their written pieces with one another. They were instructed to provide constructive feedback by offering suggestions and comments aimed at improving their peers' writing based on a set of questions. Their main role was to guide their partner by providing feedback on areas that could be enhanced, such as vocabulary usage, sentence structure, or overall coherence. This peer feedback approach encouraged collaboration and critical thinking in evaluating and improving writing.

Group B (GenAI-Assisted Revision Group) (n=18). The second half of the class (Group B) utilized GenAI to assist with their writing revision. Participants in this group had access to ChatGPT to complete the task. For the revision tasks, all students in Group B used the same pre-defined query that guided them in using the GEnAI tool effectively. All students were asked to enter the following direction in ChatGPT: «Please correct my essay below in Portuguese and explain my mistakes in English». Students were not permitted to rely on entire text outputs generated by ChatGPT. Instead, they were allowed to borrow individual words or phrases suggested by the GenAI, as well as make corrections to grammar and spelling. This approach was intended to help students focus on improving the accuracy and quality of their writing while still maintaining their own voice.

After completing the revision process, both groups submitted their revised writing. The GenAI-assisted group submitted their work electronically, including the GenAI-generated comments and suggestions. On the other hand, the peer review group submitted their revised pieces in paper format, along with the peer feedback they had received. This allowed for a direct comparison of the impact of peer feedback versus GenAI-assisted feedback on the participants' writing quality.

Post-test (Writing review for text expansion, *n*= 34)

Following the intervention phase, both groups were asked to revisit their initial writing and expand upon it - to add at least three more sentences. The goal was to evaluate the immediate changes in their writing abilities and the overall quality of their compositions after receiving feedback through different methods. Students were allowed 25 minutes to complete this task.

Peer Review Group (Group A, n=16). Participants in the peer review group helped each other revise their writing, without the assistance of digital resources, such as electronic translators, for example. They were instructed to expand their writing without external assistance, relying on the feedback they had received from their peers based on a feedback form. This allowed for a comparison between the results of the GenAI-assisted revision and peer feedback, providing insight into how each approach influenced the participants' writing skills.

GenAI Group (Group B, n=18). Participants in the GenAI-assisted group were instructed to use ChatGPT to help elaborate on their own ideas. This process allowed them to refine their writing further with GenAI offering suggestions for enhancing vocabulary, sentence structure, and overall coherence. The GenAI group focused on improving their work based on the ChatGPT's feedback, with the aim of expanding their ideas and refining their writing style. All participants entered the following direction on ChatGPT, in order to obtain consistent results: «Please make suggestions in English to improve and expand the text below».

After completing the writing revisions, both groups were asked to fill out a short survey consisting of three questions designed to assess their experience with the intervention methods. Below are the questions used in the survey with the peer review group:

1. How helpful did you find the Peer Review process?
2. Did the Peer Review suggestions improve your confidence in writing?
3. Did you notice any improvement in your writing as a result of using Peer Review?

For the GenAI group, the corresponding questions were:

1. How helpful did you find ChatGPT?
2. Did the AI suggestions improve your confidence in writing?
3. Did you notice any improvement in your writing as a result of using GenAI?

This survey aimed to gather qualitative feedback from the participants, allowing for a deeper understanding of their perceptions and experiences with GenAI and peer review. The responses provided valuable insights into how each method influenced the students' writing confidence and their perceived improvements in writing quality.

Results and Discussion

This section outlines the study's findings and provides a qualitative analysis of data collected during the pre-test, intervention phase, post-test, and subsequent survey.

1. Pre-Test Results (Baseline Assessment. First writing draft by pen and paper)

The pre-test served as a baseline assessment of participants' writing skills before the intervention. Both groups (East Coast, $n=20$ and West Coast, $n=14$) completed the writing task titled *Seu Fim de Semana Prolongado*, which aimed to assess participants' baseline writing skills. Initial analysis of the pre-test data showed that there was no significant difference between the writing abilities of the two groups, suggesting that both groups started from a similar intermediate level of proficiency. Overall, participants' pre-test contained similar inaccuracies related to gender and number agreement, verb conjugation in the past tense, spelling, use of linking words, word order, and word choice, as exemplified in the sentences below from different essays:

> Eu foi para casa onde mora minha família. (Student A)

> Fiquei em casa e descancei. (Student B)

Generally, the students' writing was marked by short paragraphs and/or a string of sentences. The intermediate level of students corresponded to the ACTFL descriptions for task and functions, context and content, text type, and accuracy (see Table 1 in this chapter). Their writing also showed their ability to create with language by combining and recombining vocabulary and language structures previously learned in class, as in this excerpt from one of the first writing drafts, written by a bilingual speaker of Spanish and English: «Bom dia! Hoje vou falar um poquinho sobre meu fim de semana prolongado. Na sexta feira eu voltei para a casa da minha família. Eu gosto de cozinhar, então eu fiz frango feito com cebola e zaatar». (Student C).

2. Intervention Phase — Review for linguistic accuracy

During the intervention phase, participants of the peer review group (Group A) took longer than the GenAI group (Group B) to review their writing. As predicted in the design of this case study, GenAI emphasizes a more individualized learning process while the peer review process is more social and collaborative. Students in the peer review took their time reading their colleague's texts and providing feedback to them. Given the fact that participants had similar levels of proficiency in Portuguese (with the exception of two heritage speakers of Brazilian Portuguese), in some instances, students could not provide help with some of the linguistics' features of the texts. On the other hand, participants in Group B received automatic feedback from ChatGPT with suggestions in English. Once students entered their typed texts on ChatGPT, it took a few seconds to receive feedback, which is exemplified below:

> Na segunda-feira escritei bastante. (Student C)
>
> ChatGPT correction: «escritei» → «escrevi» — The correct conjugation of the verb «escrever» (to write) in the past tense (pretérito perfeito) is «escrevi» for «I wrote». The form «escritei» does not exist.

Although most of the explanations of mistakes provided by ChatGPT were accurate, as in the example above, some did not capture nuances between English and Portuguese, recurring to direct translations and generating text that is not authentic in Portuguese as in the following example:

> ChatGPT correction: «Eu tinha uma noite de filme». → «Eu tive uma noite de filme». — «Tinha» is the imperfect past tense, which is used to describe ongoing actions or situations in the past. Since you're describing a complete action, you should use «tive» (preterite tense) instead.

Although the explanation above for the verb conjugation is accurate, ChatGPT did not correct the word choice. The expression «uma noite de filme» is an Anglicism in Portuguese and is not commonly used. The other downside of ChatGPT is that participants may not engage with the sentence corrections as one might expect. Since the task in the case study did not ask for students to analyze each correction, we are not certain that students have reviewed all the inaccuracies and explanations provided by ChatGPT. On the other hand, in the peer review process, students reviewed their essays in collaboration with a classmate. Based on classroom observations and on the forms filled out by students, it was clear that they engaged with each other's feedback. Students in the peer review group also spent time talking to each other and explaining how they saw each other's work. Peer review students analyzed the task completion, the comprehensibility of the texts, level of discourse, vocabulary, language control, and mechanics following the direction of the peer review form. They also provided a list of corrections for the words circled and labeled in the text. The 25-minute timeframe for this activity was sufficient to review each other's essays but insufficient to rewrite a revised version in class. Most students in Group A completed that step at home, while students using ChatGPT left the class with their final version of the essay submitted in the school's platform, Canvas. The slower pace of revision and rewriting at the intermediate L2 level has been noted in previous research.[9] There is a cognitive overload and the need for the L2 students to pay attention to several linguistic aspects while also focusing on content, slowing down the writing because of the lack of proficiency in the target language.

One common feature of both peer review and ChatGPT-assisted review is the encouraging tone of the feedback: «Your Portuguese is really good! Most of your mistakes come from interference from Spanish, which is totally normal for Spanish speakers learning Portuguese».[10]

3. Preparing to Re-write and Post-Test Results

Following the intervention, both groups were tasked with revising their initial writing and additional content to their piece. The results from the post-test indicated that both groups showed improvements in their writing in different ways, as predicted, including the ability to write creatively. Participants incorporated suggestions to expand their texts and provided more details in their narratives.

---

[9] Kormos 2024; Stevenson 2007.
[10] ChatGPT feedback for Student C.

Positive feedback was identifiable in the intervention phase, with participants from Group A writing comments, such as «This was very good, but maybe more details/ information on who you were with. Even the mundane things, like which video games you played and for how long» (Student D). A similar kind of feedback on how to improve and expand the text was provided by ChatGPT for students in Group B: «Your text is already well-structured, but you can enhance it by adding more details, transitions, and emotions. Here are some suggestions to develop it further».[11]

One noticeable trend on the ChatGPT feedback was how some of the suggestions and positive reinforcement looked very similar among students with very different writing abilities, as in: «Your text is already well-structured, but here are some suggestions to develop it further by adding more details, emotions, and reflections».[12] When compared to the ChatGPT suggestion for student F, there is little difference in feedback. However, the level of writing between Student F (Intermediate-low student, heritage speaker of Spanish) and student G (Advanced-low student, heritage speaker of Portuguese) is significant. Nonetheless, the specific suggestions provided by ChatGPT were useful and helped students to elaborate their texts further, including more details and nuances. Below is the example of ChatGPT suggestions for student F:

1. Add More Descriptions – Describe how you felt during each activity. For example, how was the party? Was the café cozy? Did you enjoy your workout?
2. Use More Connectors – Words like além disso (besides that), por isso (because of that), and enquanto isso (meanwhile) make the text flow better.
3. Expand on Activities – Instead of just mentioning studying, what exactly were you studying? Did you find it challenging?
4. Include Some Dialogue – You could mention something funny or interesting you talked about with your friends or family.[13]

Following the above suggestions from ChatGPT, Student F expanded her text by including 3 new sentences in her essay, as for example the sentence about the description of the café: «O

---

[11] ChatGPT feedback for student F.
[12] ChatGPT feedback for student G.
[13] Text generated by ChatGPT.

café novo me ajudou estudar melhor porque o ambiente de o café era tranquilo e cheio de luz» (Student F, bilingual Spanish-speaker). As exemplified in this sentence, the student does not have a full command of some basic language structures, which is expected at the intermediate level, but she is able to add more details and makes her text more developed. Student F is able to write creatively, resourcing to vocabulary not necessarily learned in class, in cognate words from Spanish such as «ambiente» and «luz» suggesting that a larger lexicon facilitates creativity in the written production as well.[14]

The quality and depth of the suggestions provided by participants in the feedback group varied among them, with some providing detailed suggestions and others making more general or limited observations. For example, Student G commented on Student's H text — «Não posso encontrar conectivos no seu texto mas há uma variedade de frases completas. Bem escrito!!». Although the text lacked cohesive devices, Student G emphasized the use of complete sentences as a positive feature. However, due to the absence of illustrative examples in this feedback, Student H did not incorporate connectors in the post-test version. In contrast, when feedback was more explicit and detailed, it had a greater impact on revision. For instance, Student I wrote to Student H «Sim, a redação reflete o vocabulário mas precisa fazer a correção dos acentos — tambem=também, quimica=química, matematica=matemática». In this case, the specific nature of the corrections, accompanied by clear examples, proved to be more effective in guiding revision. These observations align with the findings of Ellis, Loewen, and Erlam, which argue that explicit corrective feedback, such as metalinguistic explanations, is generally more effective in promoting language acquisition than implicit feedback.

On the other hand, ChatGPT is consistent in providing 3 or 4 specific suggestions and also adding vocabulary to it, such as the connectors from the example above (além disso, por isso, enquanto isso). Peer-review students generally lacked the prior knowledge to guide their colleagues, but their feedback was still valid since they provided ideas and helped students to enrich their texts as in the example below:

Suggestion 1: Expand on what time you went to the theater.

---

[14] Fernández-Fontecha 2021; Fernández-Fontecha and Kenett 2022.

Suggestion 2: Expand on which café you went after the theater and what all you bought.
Suggestion 3: Expand on how the weather was and what time of day you went to Greenlake.
Suggestion 4: Expand on what shores and homework you did on your days off. (Student K)

Although there was variation in the kinds of feedback from peer reviewers, they all made valid suggestions in terms of content expansion. In isolated cases, some suggested vocabulary variations that were not grammatically correct as in the suggestion from Student L: «Suggestion 3: Issa noite to make it flow nicer». The definite article should be «Essa» since «Issa» is not the correct spelling in Portuguese. The next session will delve into the perceptions of students from Group A and Group B regarding the kinds of feedback they received.

4. Survey

After completing the post-test, participants were asked to fill out a short survey assessing their experiences with either the peer review process (the control group) or with the GenAI tool.

4.1. Peer-Review Groups Survey

In the peer review groups, most participants found the feedback from their peers helpful, with many noting that it increased their confidence in writing. They also enjoyed the collaborative nature of the process, which made them feel more supported and engaged. In response to: «How helpful did you find the Peer Review?», students shared their realization of how they could help each other and how their own writing could improve in the process:

It was helpful because I was able to read my peer's writing and it made me realize that we are all writing at a similar level. I also liked the suggestions I was given on my writing and thought that they were all things I could improve (Student I)

I enjoyed the peer-review. I think people have a hard time realizing their own errors. So, working with a colleague, doing a peer-review helped me find my errors. (Student J)

To the question: «Did the Peer Review suggestions improve your confidence in writing?», responses included:

> Yes, it made me realize that we are all at a similar writing level. I also learned that we all make similar mistakes and they were all related to spelling, verb form and accents. (Student K)

> Yes, mainly because I know I won't make these same mistakes again in the future - in my writing. (Student L)

The response above points out to the engagement with one's own language inaccuracies and how the process of noticing mistakes and correcting them are important in language acquisition. In contrast, there are greater reservations regarding how students engage with the feedback provided by ChatGPT as a whole, as there was not any kind of activity focusing on such mistakes after they were corrected by GenAI in this case study. This should be further explored in future research.

Finally, in answering: «Did you notice any improvement in your writing as a result of using Peer Review?», participants reported:

> I noticed that I became a lot more conscious of my spelling and use of accents. I have started double checking my writing to make sure I am not missing any accents. (Student K)

> I think the peer-review has made me aware of repeated errors that I make, and I know not to repeat them in the future. (Student L)

Overall peer review feedback offers valuable opportunities for collaborative learning and peer-to-peer interaction. Some of the comments from the survey pointed to similar results in both student cohorts (Group A and Group B), such as the attention to some error patterns and the perceived help of ChatGPT and the peer review process. The interactive nature of peer reviews fosters a supportive environment that promotes the development of writing skills. However, as previously mentioned, the effectiveness of peer feedback may be limited by the language proficiency of the students involved.

4.2. GenAI Groups Survey

A large number of participants in the AI-assisted groups reported that they found the GenAI tools helpful. Several noted that the AI's suggestions improved their confidence in writing, particularly by enhancing their grammar and vocabulary.

In response to the question: «How helpful did you find ChatGPT?», students highlighted specific benefits regarding the detailed explanations and the fast correction:

> I found AI tools to be helpful because it specified which word/phrase was incorrect, the correct form of it, and the reasoning as to why it was incorrect. (Student M)

> Yes, it corrected my responses very quickly and even gave me suggestions to fix my incorrect phrases. (Student N)

When asked: «Did the AI suggestions improve your confidence in writing?», most participants responded positively, indicating that the grammar corrections and suggestions raised their confidence in writing:

> Yes, AI suggestions did improve my confidence in writing. It clarified some grammar questions that I was unsure of. (Student O)

> It did to some degree. I knew which specific sentences or phrases I was stuck on when writing the paragraph, and the AI tools helped me understand the correct way of going about using them. (Student P)

For the final question: «Did you notice any improvement in your writing as a result of using Gen-AI?», most participants mentioned positive gains:

> I did notice improvement, particularly, for words/phrases that I didn't know I was using incorrectly. I realized that my frequent errors include accent marks and feminine/masculine nouns, so I am now more conscious about it. (Student Q)

> Yes, ChatGPT found errors that I did not know. ChatGPT helped me improve the words in my text. (Student R)

The results above may also indicate a previous inclination to the use of ChatGPT as several participants had already incorporated GenAI into their academic writing and research. A few participants were more hesitant in their endorsement of GenAI, mentioning that they did not like that GenAI changed the content of their texts and their style of writing: «I found the AI tools to be a bit helpful. This is because it helped me correct grammar mistakes that I had, but at the same time it also was changing the way I write». (Student R). Another student noticed that not only the writing style changed, but the intended meaning as well: «It often misunderstood what I was trying to say and corrected for something else» (Student S). Other comments portrayed the help of ChatGPT with reservations as «moderately helpful» (Student T) and «presumptuous sometimes» (Student S). In contrast, all participants in the peer review consistently expressed satisfaction with the process.

Based on the majority of participants' feedback (n=14 out of 18 in Group B), GenAI tools may offer an efficient and personalized approach to addressing individual learners' needs. When used appropriately, these tools may help learners become aware of common, often unnoticed errors, especially the ones related to non-salient structures such as gender/number agreement, stress marks, spelling, and others. There is also the potential for GenAI to help students enhance vocabulary, grammar, and overall writing quality while also promoting creativity through suggestions and ideas to expand the text. However, further research is required to investigate the long-term effects of GenAI tools on learners' writing skills development.

Conclusion

This study explored the impact of the GenAI tool, ChatGPT, on the development of intermediate level students' writing skills in Portuguese as an Additional Language, with a particular focus on linguistic accuracy and creativity. We conducted a case study with 34 PAL students focusing on the research questions: How may ChatGPT contribute and/or hinder the students' ability to review their writing and to create with language? The study was composed of two groups: Group A — Peer review group, which functioned as a control group, and Group B — GenAI group. The results suggest that, overall, both the peer review and GenAI approaches contributed positively to students' writing development, each offering unique advantages and limitations. The peer review process fostered collaboration and deeper engagement with feedback. Students became more aware of common errors, gained confidence, and appreciated the shared learning experience. However, the peer review's effectiveness was occasionally hindered by the reviewers' own language limitations in Portuguese, especially when addressing more complex grammatical structures or suggesting diverse vocabulary. These findings go hand-in-hand with the literature on the cognitive overload of L2 writers and the attention deficiency imposed by it, especially at lower levels of proficiency when the writer lacks automaticity in the second language.[15]

In contrast to the slow process of peer review, the GenAI-assisted approach provided immediate and detailed feedback, although not always accurately. Students were able to compare their writing and incorporate GenAI suggestions. In this process, students improved their grammatical accuracy, added new vocabulary, and more details to their writing, having an impact on their creativity as well. Yet, while GenAI offered valuable suggestions, some students felt that the tool altered their writing voices or failed to consider nuanced language usage and cultural context. GenAI offers promising support for writing and creativity but must be used mindfully. The key point is designing classroom practices that empower students rather than replacing their skills. We recognize the shortcomings identified by McKee in the use of GenAI, such as distortions, violations of copyrights, homogenization of language and thinking, bias, and the risk of loss of agency in human thinking and learning. Some of these challenges may limit creativity as well.

Importantly, both groups showed progress in their post-test writing, with evidence of more elaboration and creativity adding more details to their texts. Students in both interventions expanded their writing, corrected grammatical inaccuracies, and expressed greater awareness of their errors and how to fix them. Therefore, survey responses reinforced the pedagogical value of each method, with most participants expressing satisfaction and reporting improvements in their confidence as writers.

Additionally, the findings underscore the complementary nature of AI tools and peer-review interactions in language learning. While GenAI excels in precision and feedback with examples, peer review nurtures collaborative learning and critical thinking. Future research should explore how GenAI may enhance learners' writing outcomes (without over-reliance) and support their language development through a long-term study with a larger sample size, including students' engagement with the GenAI corrections and creativity.

---

[15] Kormos 2024.

Bibliography

American Council on the Teaching of Foreign Languages (ACTFL). 2024. «*ACTFL Proficiency Guidelines Overview*». ‹https://www.actfl.org/proficiency-guidelines-overview›

Bender, Emily. 2024. «ChatGPT Has No Place in the Classroom». *Buttondown*. 22 November 2024. ‹https://buttondown.com/maiht3k/archive/chatgpt-has-no-place-in-the-classroom›

Bender, Emily, Timnit Gebru, Angelina McMillan-Major and Shmargaret Smitchell. 2021. «On the Dangers of Stochastic Parrots: Can Language Models Be Too Big?». *FAccT '21: Proceedings of the 2021 ACM Conference on Fairness, Accountability, and Transparency*. New York, Association for Computing Machinery. 610-623.

Bedington, Andelyn, Emma Halcomb, Heidi McKee, Thomas Sargent and Adler Smith. 2024. «Writing with Generative AI and Human-Machine Teaming: Insights and Recommendations from Faculty and Students». *Computers and Composition* 71: 1-13.

Chenoweth, N. Ann and John Richard Hayes. 2001. «Fluency in Writing: Generating Text in L1 and L2». *Written Communication* 18.1: 80-98.

Fernández-Fontecha, Almudena. 2021. «The Role of Learner Creativity in L2 Semantic Fluency. An Exploratory Study». *System* 103: 1-14. ‹https://www.sciencedirect.com/science/article/pii/S0346251X21002128›

Fernández-Fontecha, Almudena and Yoed Kenett. 2022. «Examining the Relations Between Semantic Memory Structure and Creativity in Second Language». *Thinking Skills and Creativity* 45: 1-13. ‹https://cognitive-complexity.net.technion.ac.il/files/2022/10/Fernandez-Fontecha-and-Kenett-2022.pdf›

Jamieson, Sandra. 2022. «The AI "Crisis" and a (Re)turn to Pedagogy». *Composition Studies* 50.3: 153-157.

Jones, Rodney, ed. 2015. *The Routledge Handbook of Language and Creativity*. New York, Routledge.

Kormos, Judit, Shungo Suzuki and Olena Rossi. 2024. «The Role of Creativity in Second Language Writing Performance». *Learning and Individual Differences* 114: 1-11. ‹https://www.sciencedirect.com/science/article/pii/S1041608024000931?via%3Dihub›

Ellis, Rod, Shawn Loewen and Rosemary Erlam. 2006. «Implicit and Explicit Corrective Feedback and the Acquisition of L2 Grammar». *Studies in Second Language Acquisition* 28.2: 339-368.

McKee, Heidi. 2025. «GenAI & Writing Across the Curriculum». Presentation at the University of Washington. ‹https:// english.washington.edu/news/2025/05/19/english-pro fessors-and-simpson-center-bring-vital-ai-conversa tions-uw›

Selber, Stuart. 2004. *Multiliteracies for a digital age.* Carbondale, Southern Illinois University Press.

Stevenson, Marie, Rob Schoonen and Kees De Glopper. 2007. «Inhibition or Compensation? A Multidimensional Comparison of Reading Processes in Dutch and English». *Language Learning* 57.1: 115-154.

Relationality as a Pedagogical Imperative in the Era of GenAI

Karina Lissette Cespedes & James R. Paradiso

It is undeniable that GenAI has completely transformed the teaching and learning landscape. Within the global north, the turn towards GenAI in higher education has raised a number of important questions about creativity, relationality, and the effects on longstanding traditional approaches to learning and pedagogy. Recognizing that the GenAI revolution is felt differently depending on the geographical region of utilization, our observations speak to the experiences of educators working in parts of the world and at institutions where access to technology and connectivity is ubiquitous, and where leveraging GenAI tools to enhance job-related productivity and instructional practice is overwhelmingly supported.[1] As faculty on one of the largest and well-resourced campuses in the southern U.S., we have been fascinated by the impact of GenAI, in particular the effects of what can be described as a divide between pedagogues that have enthusiastically leaned into the new normal, and those who openly vocalize concerns about the negative impacts of GenAI on teaching and learning.

On GenAI Transforming Education: Looking into the Machine

Without a doubt, large language models (LLMs) are convenient for several natural language processing (NLP) tasks, as evidenced by the proliferation of LLM-powered generative artificial intelligence (GenAI) tools that are revolutionizing the way societies with access to this technology interact with information. Institutions of higher learning have adapted quickly to the incorporation of GenAI, and faculty, rushing to accommodate the rise of GenAI, have begun the process of retraining. According to a national survey of postsecondary faculty conducted by Reudiger, of roughly 2,600 responses, approximately fifty percent (50%) of instructors reported seeing value in receiving instructional support related to GenAI. This data has bolstered institutional confidence, motivated training

---

[1] See Chen 2023; McDonald 2025.

and support units within the academy to allocate sufficient resources to meet the technical, pedagogical, and ethical needs of instructors, as they seek to support student success in the middle of a GenAI revolution.[2] Part of the promise of receiving support, as faculty train on GenAI, has been captured by Karpouzis, who posit that through intentional prompt engineering (i.e., dialoguing), GenAI can help instructors create meaningful, customized lesson plans that accommodate a variety of learner profiles.

However, just as interesting have been the discussions among concerned pedagogues about the rise of GenAI, many of whom have questioned the fast implementation of the new technology within higher education, the expediency with which information is acquired and the mimicry of human creativity as problematic for the future of student learning, development of ingenuity, retention and application of information. Concerned pedagogues point to the folly of GenAI «hallucinations» or fail rates, which can vary from 83% (DeepSeek) to 62% with OpenAI, as misleading or misinforming students. They depict a dystopian conundrum in which the potential outcome is a post-literate society emerging in the wake of GenAI, unable to fully formulate convictions born out of the complex praxis of human thought and experimentation. In the article *Trusting the Machine: How Generative AI is Reshaping Critical Thinking and Knowledge Networks*,[3] author Joal Baum takes on such concerns by mapping a complex shift from human validation of information to GenAI mediated knowledge. Here, Baum, drawing on the scholarship of Lee, *The Impact of Generative AI on Critical Thinking* incorporates data on human cognitive effort and highlights the finding that GenAI reduces the effort of knowledge retrieval and creation of content, in short, diminishing critical thinking. This occurs in a context in which high trust is placed on GenAI, according to Baum and Lee, where there is less scrutiny and evaluation based on the lived experiences or expertise of the user. But the study also finds that in a different context, particularly those characterized by high user expertise on a topic and/or prior information confidence, there remains greater skepticism of the GenAI re-

---

[2] See Leslie and Meng 2024; Sahu 2024.
[3] Baum 2025.

sponses. Lee points to this difference as an important indication that GenAI, more than about increasing productivity, can become a useful tool through which to assess how knowledge is produced, when it is validated, and trusted. A key takeaway from the study is that the process of validation and trust born out of human experiences and expertise are key components in defining the role humans play. In deliberating the dynamic debate between enthusiastic and concerned pedagogues, it appears that at the core of the two perspectives sits a deeper question about the use-value of learning and information sharing as an endeavor, as a pro-social human right and enterprise. The dynamic debate we are witnessing becomes a way to ponder the following: Is the current impact of GenAI unique or novel? Have there been earlier epochs that have transformed learning, work, and education in such a way that it ushered in a type of pre-GenAI «artificial intelligence»? Also, regardless of GenAI, how can the pro-social merits of education grow?

The Knowledge Economy and «Artificial Intelligence» Before GenAI

Current debates on GenAI taking place within societies that enjoy high rates of connectivity, awash in well-established electrical grid infrastructure such as the U.S., are occurring within the same geo-political locations that experienced the powerful birth of the knowledge economy after WWII. By the mid-1960s, the projected main industry of nations booming after the war gradually emerged as a robust information and service industry slated to replace manufacturing. In the face of this historical mid-twentieth-century shift, some may retort to both the GenAI enthusiasts and those concerned about its effects that the debates we witness today are a continuation and evolution of this post-WWII dynamic social and economic transformation. In many ways both the enthusiasm and trepidation expressed currently is but this era's version of the hopes and fears expressed by previous generations — some heralding a new world where hard physical labor would be eased, others lamenting, for example, the rise of industrialization, then mechanization, followed by the mid-twentieth century concerns over the rise of the knowledge economy in the wake of deindustrialization. We find it compelling to witness how the current restructuring of the information economy, as

it has been transformed by corporations in the U.S. and resulted in the quick adjustments made within higher education,
has induced many of the same dialectical concerns that previously motivated philosophers, students, pedagogues and
workers during the middle of the last century. This time, however, the concern over the new synthesis, the end of the hu-
man-centered knowledge economy, merges two fears — that
of hyper-technological mechanization and the end of a world
order sustained by human acquisition of information at the
center of the knowledge economy. Moreover, it is interesting
to consider how a mid-twentieth-century fear of the rise of a
knowledge economy threatening traditional learning and or-
ganized labor has been inverted into today's trepidation over
the GenAI decimation of the same. As two pedagogues witnessing these dynamic debates and their projections of massive
shifts in labor and learning, we find ourselves walking the
rocky interstitial space between various positions. And, we
have found ourselves musing deeply on what the deliberations
on GenAI could mean for student learning, creativity, and the
role of relationality. We take seriously all positions on the
emergence of GenAI — the optimistic perspective that has
spearheaded the development of pedagogical approaches incorporating GenAI into student learning, as well as its counter
which include a range of warnings — from calls to question
the trust some have place on GenAI (Baum), to approaches ex-
uding a metaphorical «rage against the machine».

Capturing the complexity of these debates, it seemed apro-
pos then to incorporate a non-GenAI created story, a vignette,
that interweaves the use of GenAI with, interestingly enough,
GenAI created music in the tradition of the renowned countercultural punk rock, metal, rap group Rage Against the Machine.

From GenAI Inquiry to Rage Against the Machine

The renowned band from LA, whose music has for decades
captured the imagination of fans interested in larger historical
and social questions, is famous for offering critiques of hegem-
ony and is itself an example of raw creativity. Due to its coun-
terculture reputation Rage Against the Machine was utilized as
a topic of inquiry by the writer Jacob Uitti. The group was a
prominent_component of his thought experiment, aiming to
test the limits of GenAI. In the article, Uitti describes the band

as «one of the most impactful, hard-pounding groups ever [...] [and] [...] a lightning bolt».[4] Originally marveling why a self-declared fan of the group would turn to GenAI to create a facsimile of a song by the iconic countercultural group it then became clear to us that the endeavor by Uitti provided an exemplary snapshot into the polemic, while signaling one of the many ways in which GenAI is not, as of yet, replacing human creativity, trust, validation and relationality.

In his article Uitti provides a personal explanation about why he nostalgically turned towards GenAI after missing the energy and creative music of Rage Against the Machine adding that the band's departure from recording and touring signaled that this had produced feelings of loss and a desire to relive, even if «in-authentically», a glimmer of the collective energy transmitted by the band and their iconic lyrics.

Uitti admits that although part of the allure of the group is its mystery, irreverence, and unpredictability, which justifies their long absence from touring, he was curious to see whether GenAI had acquired enough data on the band's unconventional internal logic to create a version of their music. The implicit question Uitti posited was, has GenAI absorbed the logic of «resistance», or would GenAI falter at the request for a song in the style of Rage Against the Machine? The question posed to GenAI presumed that GenAI, which has proven capable of replicating the lyrics of more mainstream popular music, would fail. In short, Uitti's request is noteworthy precisely because it demonstrates a personal contradiction — on the one hand the writer is eager to nostalgically relive the thought provoking lyrics the band is most known for, yet curious to see if the power of Rage Against the Machine can, well, actually rage against GenAI potentially becoming confounded by the anti-establishment logic to such an extent that as a fan Uitti would be able to recognize, to validate, the difference between what is real and what is at best a simulacrum.

The GenAI output resulted in the song titled *A Fury*[5] that, according to the writer, impressively held to standards fans have come to expect.

---

[4] Uitti 2023.
[5] Uitti 2023.

What Uitti discovered via this thought experiment is that indeed GenAI can yield a composite of the counter-culture perspective of Rage Against the Machine based on an amalgamation of the band's lyrical repertoire. Essentially, identifying patterns, offering inductive statistical probabilities, this led Uitti to conclude his article by stating he was «surprised that ChatGPT could go this hard [and] [...] gets right to it [...] didn't necessarily think that a machine would channel Rage Against the Machine this well».[6]

Yet, Uitti's response yields more than praise for the GenAI lyrics. His observations affirm his own expertise to validate the output as «convincing». Moreover, just as quickly as Uitti admits that the lyrics «get right to it» he notes, with a tone of disappointment that the GenAI created *A Fury* failed to fully quell his nostalgia by concluding, «hope the band will release more new music — it would be a welcomed sight [...] But until then [...] we have the [GenAI lyrics] to keep us occupied».[7]

Uitti's validation that GenAI's *A Fury* can «keep us occupied» isn't the end of the story, but only the beginning. Pushed further, the GenAI produced *A Fury* signals the possibility of a relational opportunity which, if Uitti would like to embark upon such a journey, could lead to engaging not with GenAI but with the living fans as well as potentially the 'real' band Rage Against the Machine to confirm whether they would similarly find the GenAI lyrics convincingly channeling the signature logic. Such efforts would potentially lead to a vigorous reconnection and sociality that is based not on a commodity, but instead sociality grounded on the ideas expressed via the lyrics, the band's philosophical position, and yield yet one more human-produced article for Uitti, which dives right into the heart of the polemic.

What Can be Learned from *A Fury*

In assessing the desire for sociality within Uitti's conclusion, three aspects of human creativity, well beyond the production of a commodity, were evident. These three markers of human ingenuity are difficult for GenAI (as of yet) to replicate, and, ultimately may not be the goal of the technology; the first, human conviction and discernment about the GenAI output,

---

[6] Uitti 2023.
[7] Uitti 2023.

captured by Uitti's statement that although *A Fury* «gets right to it» it doesn't fully get it right, which is why he still hopes «the band will release more new music»[8] as it would be a welcomed sight. Second, the pull of relationality as expressed by Uitti in the form of the original and lingering nostalgia for the band's music, its touring, and the sense of human connection provided by concerts, led Uitti to request a speculative GenAI song in the first place. And, Uitti's own vast expertise as a longtime fan and writer conveys that, in the end, GenAI created an approximation, merely something to «keeps us occupied», until the «real» band produces new content. It signals that the conversation being mediated via GenAI is still very much between humans.

Uitti's experimentation and expertise in crafting his article's analysis provides an example for pedagogues, and a path for converting GenAI angst and «rage against the machine» into student learning opportunities within the humanities. By centering the learner's/ user's expertise and showcasing personal areas of interest, GenAI can serve the goal of connecting with others and expanding human understanding. To this end, some educational practitioners, for instance, have begun moving away from assignments and assessments that can be easily completed with the assistance of GenAI, opting instead to measure students' knowledge/content proficiency through their production of novel, creative works that prove challenging to generate through prompting a GenAI system.

Such endeavors can foster an exploratory learning environment where human creativity, conviction and knowledge expands on various fronts as it relies on learners drawing upon their own understanding, convictions and definitions of what they consider worthwhile and significant application of information thereby pushing learning beyond the rote regurgitation of information, or what the pedagogue Paulo Freire described as the banking concept of education. Here, Freire's theorizations on the banking concept of education provide a provocative lens through which to consider how, from a Freirean point of view, the rise of «artificial intelligence» pre-dated GenAI within educational settings. Freire's theorization would take into consideration how Twentieth and Twenty-First century education models that prioritized rote memorization or

---

[8] Uitti 2023.

formulaic writing with the goal of producing a workforce in the service of capital would be incentivized to showcase the «downloading» of information, without conviction or positionality had long engendered its own forms of «artificial intelligence» — producing learning styles that are functionally literate, capable of memorization and recall but lacking dynamic synergy, inquiry, authority to assess information based on lived knowledge. Moreover, this type of banking concept of education and pre-GenAI version of a non-technologically based «artificial intelligence», stifled robust collective learning, creativity, and relationality. The litmus test for such an environment could be found in the stilted writings of students and the inability of multiple-choice exams to convey depth of understanding. As such, the difficulties that some pedagogues today warn are on the horizon due to GenAI may have very well been at play long before the rise of GenAI, and would have potentially continued without the rise of GenAI.

In response to the banking concept of education, Freire proposed a pedagogical model theorized as «conscientization», which emphasized dialogue and communication as a method of developing what the Philosopher Hanna Arendt described within her writings at the end of WWII as the role of conviction in education. For Arendt, actively fostering the development of convictions as a goal of education ultimately makes it difficult to destroy the capacity to form any. This process of fostering conviction and engaging in dialogue was challenging to secure long before GenAI, as standardized testing, rote memorization, or writing prompts that required the regurgitation of information, versus the application of concepts to convictions, observations, and lived experiences, were normalized within higher education for the sake of expediency. From the theoretical perspective of Arendt and Freire, human inquiry and conviction were not a guarantee within education systems, and what we are witnessing is a potential swapping out of the «artificial intelligence» from information held by humans exchanging their labor within the knowledge economy to what GenAI can expeditiously provide to industry as an output. In the end, regardless of which system of labor and education dominates, what remains for humans is what has always been the core of humanity — conviction, creativity, and expertise.

It is important to note that we, as two pedagogues at a large U.S. institution, are well cognizant of the myriad pressures on faculty and learners, which make the challenges some projects will encounter with the ubiquitous use of GenAI impossible to ignore. We are also just as aware that implementing Freire's theorization requires a reframing. As a part of a reframing, Freire theorized the «banking concept of education's» inter-connectedness with employment as important to keep in mind. Fast forward to the contemporary learning environment, the concerns expressed by pedagogues stress that the technology expeditiously replaces critical thinking under pressure to meet short-term pragmatic labor or education tasks, propelling users, within a context of speedup, to engage with GenAI. For learners, this can be the pressure to finish an assignment while facing a dreaded deadline. In such instances, the urgency tied to assignment generation and test-taking designed under a different educational regime to assess retention of information becomes unreliable for pedagogues across the country, already stretched thin by increased workloads and a large number of students. Such realities have spawned multiple «how-to-teach with GenAI» conferences, accelerated the growth of GenAI tutors, teaching assistants, and a myriad of publications at the intersection of education and GenAI. Pedagogues, walking the proverbial tightrope between the rise of GenAI and the need to assess student progress within the banking concept of education, have earnestly compiled collections of GenAI assignments. All efforts, however, point to the same conclusion: they by default acknowledge the elephant in the classroom, which is that traditional assessments are inadequate. And, that these same traditional assessments may have always been less than reliable in gauging learning.[9]

Applying the theorization by Freire on the banking concept of education would suggest that traditional assignments prior to GenAI were not a guarantee that learning assessments were functioning. It would then posit the hypothesis that prior to the rise of GenAI, multiple choice assignments and rote memorization failed to function as indicators of learning, mastery of subject. This understanding has been supported by research on the ways in which rote memorization has long been known

---

[9] Brookhart 2017.

to not result in long-term retention and application of information.[10] Freire's theorization would expand the above observation.

However, when personal inquiry and exploration is factored into learning environments, including environments under pragmatic pressures to meet the demands of industry, the learning experience can be transformed. As such, by re-imagining the ways in which GenAI can be deployed to foster engagement between learners and pedagogues, the relational significance of learning, and the ways in which learning occurs long before any one assignment is due, can yield new opportunities.

Acknowledging that GenAI will likely not become obsolete in the near future, we look to identify methods that build relational pedagogical approaches, incentivize ethical use of GenAI, while also navigating pragmatic concerns about the impact of GenAI on knowledge production and potential restructuration in which GenAI has emerged as both a resource tool and as a labor competitor. With this understanding, establishing certain safeguards may be helpful to consider when incorporating GenAI tools into existing pedagogical standards. As well as asking the following question: How might instructional methods need fine-tuning to accommodate new conditions being introduced to current (and future) educational systems? How might students be encouraged to use GenAI ethically? How might educators feel empowered, rather than terrified, of a future with highly sophisticated, autonomous tutoring systems and bot essay writers with impeccable grammar and structure, yet questionable content expertise with numerous inherent biases and an inability for (inherent) sensemaking? No matter the question or context within which it is asked, instituting clear guidelines on the ethical use of GenAI in course design and development, content creation, as well as the implementation and evaluation of instructional methods as they relate to student success, is imperative.[11] The value formed through such convictions and thoughtful consideration can then permeate institutional policies and practices, ensuring consistency and clear expectations for all who may be im-

---

[10] Lee 2025.
[11] Tillmans 2025.

pacted. The hope, of course, is that these 'guidelines' are established dialogically through interactions with students and with as many stakeholders as possible to mediate how GenAI technologies evolve much faster than institutional policies, leaving a potential gap or misalignment between institutional practices and the end user expectation and/or experience. Therefore, supporting students and faculty by helping them navigate critical (timely) matters, such as instructional content production (in the case of faculty) and academic assignment and assessment completion (in the case of students).

It is helpful to consider highly-contextualized learning conditions (e.g., background knowledge, resource exposure), where the notion of personalized learning is highly attractive,[12] and if GenAI can support this endeavor,[13] by centering the role of student discernment, conviction, creativity and expertise. Additionally, as not all faculty members express high levels of enthusiasm, optimism, or comfortability using GenAI,[14] another path would strongly signal the emergence of an opportunity for professional development/training that could empower both pedagogues and students while ensuring dialogue and relationality.

Navigating the «new normal»: Recommendations for Using GenAI in Educational Settings

At this point it is clear that many pedagogues have experienced the need to adapt, to prepare for the practical, theoretical, and ethical impacts of GenAI in postsecondary teaching and learning contexts. Implementing any recommendation and strategies, however, must take into consideration multiple variables at play when looking to upskill instructors: As Zheng noted, while technological knowledge is undeniably important, effectively utilizing GenAI involves combining technical comfortability with pedagogical know-how and a certain level of adeptness regarding ethical standards and their implications in education.

Based on the work of Cordero, Luo, and Symeou, the following list, compiled by Jim Paradiso, offers some suggestions:

---

[12] Zheng 2022.
[13] Hoernig 2024; Khan 2024.
[14] See Ceballos 2024; Crompton and Burke 2024; Toncelli and Kostka 2024.

o   Explore how students are using GenAI as well as the limitations and affordances of different GenAI tools. (Note: Knowing how students use GenAI tools for learning is just as important as knowing how to use the tools within teaching.)

o   Revise assignments and assessments (as appropriate) in ways that take the features of GenAI tools into consideration.

o   Have students actively engage with GenAI tools through open-ended problems and creative writing tasks. Similar to the ways in which Uitti assessed the capabilities of GenAI to mimic the logic of the band Rage Against the Machine.

o   Promote critical thinking and creativity, being aware that student reliance on GenAI to solve problems or craft textual content may result in a decline of these skills. Harness the expertise students bring with them into the course.

o   Institute GenAI policies at the course start and reiterate pertinent aspects of these policies any time the tool might be leveraged to complete a task (e.g., graded assignments, discussions).

o   Consider «originality» across a spectrum, accounting for any variety of collaborative dialogue between students and GenAI and emphasizing distinctions between tasks at which GenAI excels (e.g., conducting large-scale data analysis) versus those at which humans excel (e.g., providing nuanced contextual insights).

o   Be aware of GenAI tools' data handling practices to ensure the security of sensitive and/or personal information.

o   And, when it comes to GenAI bias, openly acknowledge and facilitate the creation of course content that is accessible for students with disabilities and sensitive to the varying initial conditions (e.g., social, educational, cultural) of students.

In addition to these practical and ethical considerations, we also consider the creative power of relational reflexivity as an aspect of pedagogy and knowledge production not replicable (as of yet) through GenAI.

While Freire developed theoretical frameworks to address the relationship between educators and learners, with the goal of active inquiry and curiosity-led participation as key to engaged learning, bell hooks offered insights into the ways in which a commitment to teaching extends beyond the subject matter and is a commitment to service.[15] Both Freire and hooks provided alternative approaches to traditional prompts and assessments. They encouraged exploration into how topics of study influence the student's understanding of the world as active learners, and hooks saw teaching as a commitment to

---

[15] hooks 2014.

service, not as a method of checks and balances, assuring the content was «downloaded» and efficiently «retrieved» on tests.

Enhancing Student Engagement Through Relational Reflexivity in the Era of GenAI

As Paulo Freire persuasively theorized, active learners read the world, not just words on a page. If we adapt Freire's perspective to the contemporary moment, then active learners do not just take at face value the response of a GenAI prompt. Operating from the perspective that humans are inherently active learners within and outside of instructional settings then it would be possible that active learners will validate information based on complex lived expertise, while seeking to provide solutions to real-world needs — whether those needs are to meet a nostalgic yearning for their favorite band, or to solve a concern about wastewater in their community. In the hopes of engaging with active learners, relational reflexivity, as a philosophical approach to pedagogy, may be of interest as it reinforces the creativity with which learners read the world via a pro-social reciprocal process, providing connectedness to self and others.

At its core, relational reflexivity gauges how a project is conceived as a significant human endeavor. Towards this end, relational reflexivity encourages active learners to enter into a dialogic relationship with ideas, peers, and faculty facilitated through the course content material, and the sharing of the final version of an assignment via a co-created publication amassing the project, for example, via an OER platform. The assignment treats the class (whether synchronous or asynchronous) as a living community interested in communicating what was learned, how it was learned, alongside the needs discovered by the learner. The active sharing of these bolsters student understanding of concepts and facilitates the creation of an applied analysis that merges course content, community perspective, student expertise and the discovery of a pro-social solution.

One approach that has produced thoughtful engagement begins by introducing learners to at least one established social science qualitative research method through which connection to a topic and to the community serves to build applied mastery of course content via original analysis. The pairing of course material and the results of qualitative research is, as of yet, versatile and nuanced enough to bypass GenAI prompts. And, for institutions committed to teaching with GenAI, it wouldn't bar students from engaging with GenAI for the gathering of data or information about a topic. An example of an assignment aiming to foster relational reflexivity within courses taught by Karina introduces students to the qualitative research method, PhotoVoice. PhotoVoice has been woven into several fields and academic disciplines for its versatility and engagement with community.[16] This method was developed by Caroline Wang and Mary Burris with the aim of combining Paulo Freire's notion of «advanced literacy» with an emphasis on the importance of voice, photography, and documenting community engagement, interests, and needs.[17] The PhotoVoice assignment can be utilized within synchronous and asynchronous courses, and facilitates shifting the conversation away from the conundrum of utilizing GenAI, towards a consideration of the significance of lived experiences and social connections, with or without GenAI as part of the query. The assignment offers students the opportunity to answer a series of questions that invite engagement with topics that are significant to the learner and to a member of their community. Learners extend an open-ended question to a member of their community, who in turn contributes images and statements that chronicle their lived experiences. Then, an analysis is crafted with the aim of exploring the identified topics via the course material and_the experiences of communities. Bolstered by the organizing principle that students bring a wealth of knowledge into the learning environment, and drawing on relational reflexivity, students engaging with PhotoVoice are encouraged to identify personally significant research topics. Through the creation of openly licensed content using digital tools, such as Pressbooks, and qualitative research methods, such as PhotoVoice, the assessment fosters an understanding of applied humanities, while building sociality around a topic of significance to the learner and their community.

It is important to note that the assignment, if needed, could be adapted to the era of GenAI, while maintaining the objective of fostering engaged literacy as an important step for students transitioning from passive consumers to producers of

---

[16] Dmello and Kras 2021; Jehangir 2022.
[17] Wang and Burris 1997.

knowledge.[18] Via such an approach, learners can creatively experiment with shifting the emphasis away from regurgitating information and instead develop an analysis that builds on academic and experiential sources, which would be shared with peers at different stages. The goal of strengthening skill sets for students via engagement with qualitative research empowers students to draw on their existing expertise, acquire new understandings, and engage with their communities meaningfully. The method of PhotoVoice is divided into a series of steps for students to follow. At the end of these steps, students emerge conversant in a community's needs and can express convictions held by a community member. Ultimately, students aim to master «the relationship between our knowledge and our practice: how we engage, critique and test ideas and theories in practice, and upon what basis we make judgments».[19] A relational approach enables engaging with topics covered across disciplines via the qualitative analysis of personal experience. For example, a project on the topic of water pollution would entail: 1) assessing the science, 2) gauging public perspective, and 3) testing existing ideas in practice. The PhotoVoice project, created by an active learner in Karina's course, centers the issue of wastewater in Brazil and provides an example of an assignment accomplishing these three components: «Sewage: A Problem of a Developing Country».[20]

Conclusion

The «innovation economy» in the age of GenAI calls on pedagogues and learners to collaborate across mutual areas of expertise. The goal, regardless of the technology, continues unchanged — the development of pro-social creativity, curiosity, courage, compassion, and communication as key to connectedness and an existence that fosters relationality. As the rise of GenAI continues to expand, and the role of educators creating thought-provoking assignments has become ever more significant, we are reminded of what philosopher and educator Hannah Arendt posited in her writings at the end of WWII. Arendt expressed that actively fostering the development of convictions as a goal of education ultimately makes it difficult to

---

[18] Freire and Macedo 2005.
[19] Dyke 2015, 549.
[20] Britto 2021.

destroy the capacity to form any. To this end, we propose an understanding of relationality that takes into consideration a new era, a new beginning in which, with or without GenAI, sharing information with others shapes what humans are able to access and how students experience the world. Not only does fostering relationality in the crafting of assignments engage students, but it can promote expertise, sociality, empathy, and the ability to validate the applicability of newly acquired knowledge to other areas of inquiry.

Bibliography

Baum, Joel. 2025. «Trusting the Machine: How Generative AI is Reshaping Critical Thinking and Knowledge Networks». *Medium*. 12 February 2025. ‹https://medium.com/data-science-collective/trusting-the-machine-how-generative-ai-is-reshaping-critical-thinking-and-knowledge-networks-7d1ae4772e45›

Britto, Marcella. 2021. «Sewage: A Problem of a Developing Country». In *From the Bottom Up*. Karina Cespedes, ed. Orlando, University of Central Florida.

Brookhart, Susan M. 2017. «The Use of Teacher Judgement for Summative Assessment in the USA». In *International Teacher Judgement Practices*. London, Routledge. 79-100

Cabellos, Beatriz, Carlos De Aldama and Juan-Ignacio Pozo. 2024. «University Teachers' Beliefs about the Use of Generative Artificial Intelligence for Teaching and Learning». *Frontiers in Psychology* 15: 1-15.

Cespedes, Karina Lissette and James R. Paradiso. 2023. «Combining Open Pedagogy and Undergraduate Qualitative Research: PhotoVoice as Method Toward Relational Reflexivity». *Journal of Open Educational Resources in Higher Education* 2.1: 269-276.

Chen, Baiyun, Aimee Denoyelles, Tim Brown and Ryan Seilhamer. 2023. «The Evolving Landscape of Students' Mobile Learning Practices in Higher Education». *Educause Review*. 25 January 2023. ‹https://er.educause.edu/articles/2023/1/the-evolving-landscape-of-students-mobile-learning-practices-in-higher-education›

Cordero, Jorge, Jonathan Torres-Zambrano and Alison Cordero-Castillo. 2024. «Integration of Generative Artificial Intelligence in Higher Education: Best Practices». *Education Sciences* 15.1: 32.

Crompton, Helen and Diane Burke. 2024. «The Educational Affordances and Challenges of ChatGPT: State of the Field». *TechTrends* 68.2: 380-392.

Dmello, Jared R. and Kimberly R. Kras. 2021. «Photovoice as a Research Technique for Student Learning and Empowerment: A Case Study from a South Texas Border Town». *Journal of Criminal Justice Education* 32.4: 513-536.

Dyke, Martin. 2015. «Reconceptualising Learning as a Form of Relational Reflexivity». *British Journal of Sociology of Education* 36:4: 542-557.

Freire, Paulo. 2018. «The Banking Concept of Education». In *Thinking about Schools: A Foundations of Education Reader*. Eleanor Blair Hilty, ed. London, Routledge. 117-127.

Freire, Paulo and Donaldo Macedo. 2005. *Literacy: Reading the Word and the World*. London, Routledge.

Hoernig, Steffen, André Ilharco, Paulo Trigo Pereira and Regina Pereira. 2024. «Generative AI and Higher Education: Challenges and Opportunities». *Institute of Public Policy*. ‹https://www.ipp-jcs.org/wp-content/uploads/2024/09/Report-AI-in-Higher-Education-IPP-1.pdf›

hooks, Bell. 2014. *Teaching to Transgress*. London, Routledge.

Jehangir, Rashné R., Kelly Collins and Terra Molengraff. 2024. «Class Matters: Employing Photovoice with First-Generation Poor and Working-Class College Students as a Lens on Intersecting Identities». *Journal of Diversity in Higher Education* 17.3: 319.

Karpouzis, Kostas, Dimitris Pantazatos, Joanna Taouki and Kalliopi Meli. 2024. «Tailoring Education with GenAI: A New Horizon in Lesson Planning». *2024 IEEE Global Engineering Education Conference (EDUCON)*: 1-10. ‹doi: 10.1109/EDUCON60312.2024.10578690›

Khan, Rubaina, Sreyoshi Bhaduri, Tammy Mackenzie, Animesh Paul, Sankalp KJ and Indrani Sen. 2024. «Path to Personalization: A Systematic Review of GenAI in Engineering Education». *KDD AI4Edu Workshop*, ACM, Barcelona, Spain. ‹https://hal.science/hal-04673700›

Kovach, Margaret. 2010. «Conversation Method in Indigenous Research». *First Peoples Child & Family Review* 5.1: 40-48.

Lee, Hao-Ping, Advait Sarkar, Lev Tankelevitch, Ian Drosos, Sean Rintel, Richard Banks and Nicholas Wilson. 2025. «The Impact of Generative AI on Critical Thinking: Self-Reported Reductions in Cognitive Effort and Confidence Effects from a Survey of Knowledge Workers». *Proceedings of the 2025 CHI Conference on Human Factors in Computing Systems*. Article No.: 1121, Pp. 1-22. ‹https://doi.org/10.1145/3706598.3713778›

Leslie, David and Xiao-Li Meng. 2024. «Future Shock: Grappling with the Generative AI Revolution». *Harvard Data Science Review* 5: 1-34.

Luo, Jiahui. 2024. «A Critical Review of GenAI Policies in Higher Education Assessment: A Call to Reconsider the "Originality" of Students' Work». *Assessment & Evaluation in Higher Education* 49.5: 651-664.

McDonald, Nora, Aditya Johri, Areej Ali and Aayushi Hingle Collier. 2025. «Generative Artificial Intelligence in Higher Education: Evidence from an Analysis of Institutional Policies and Guidelines». *Computers in Human Behavior: Artificial Humans* 3: 101-121.

Ruediger, Dylan, Melissa Blankstein and Sage Love. 2024. «Generative AI and Postsecondary Instructional Practices». *Ithaka S+R*. June 20. 1-17. ‹https://doi.org/10.18665/sr.320892›

Sahu, Ms Neetu. 2024. «The GenAI Revolution: Unleashing the Role of Information Technology in Education». *Sudarshan Research Journal* 2.5: 55-57.

Seraphin, Sally B., J. Alex Grizzell, Anastasia Kerr-German, Marjorie A. Perkins, Patrick R. Grzanka and Erin E. Hardin. 2019. «A Conceptual Framework for Non-Disposable Assignments: Inspiring Implementation, Innovation, and Research». *Psychology Learning & Teaching* 18.1: 84-97.

Symeou, Loizos, Loucas Louca, Argyro Kavadella, James Mackay, Yianna Danidou and Violetta Raffay. 2025. «Development of Evidence-Based Guidelines for the Integration of Generative AI in University Education Through a Multidisciplinary, Consensus-Based Approach». *European Journal of Dental Education.* ‹https://pubmed.ncbi.nlm.nih.gov/39949032›

Tillmanns, Tanja, Alfredo Salomão Filho, Susmita Rudra, Peter Weber, Julia Dawitz, Emma Wiersma, Dovile Dudenaite and Sally Reynolds. 2025. «Mapping Tomorrow's Teaching and Learning Spaces: a Systematic Review on GenAI in Higher Education». *Trends in Higher Education* 4.1: 2.

Toncelli, Rachel and Ilka Kostka. 2024. «A Love-Hate Relationship: Exploring Faculty Attitudes Towards GenAI and Its Integration into Teaching». *International Journal of TESOL Studies* 6.3: 77-94.

Uitti, Jacob. 2023. «AI Writes a New Rage Against the Machine Song». *American Song Writer,* November 2023. ‹https://americansongwriter.com/ai-writes-a-new-rage-against-the-machine-song›

Wang, Caroline and Mary Ann Burris. 1997. «Photovoice: Concept, Methodology, and Use for Participatory Needs Assessment». *Health Education & Behavior* 24.3: 369-387.

Zheng, Lanqin, Miaolang Long, Lu Zhong, and Juliana Fosua Gyasi. 2022. «The Effectiveness of Technology-Facilitated Personalized Learning on Learning Achievements and Learning Perceptions: A Meta-Analysis». *Education and Information Technologies* 27.8: 807-830.

Devising with the Machine: Artificial Intelligence and Human Creativity in Performance

Chloë Rae Edmonson, Julia Listengarten & Alyssa Barrack

As theatre artists and scholars, we believe the performing arts are uniquely positioned to explore what it means to be human in an era defined in part by the acceleration of Artificial Intelligence (AI). On one hand, the interaction between human performers and audience members in a live theatre space remains irrevocably embodied; on the other, the advent of AI offers unique opportunities for integrating dynamic technologies into the fabric of our performances. This chapter recounts our experience creating an experimental theatre performance for the UCF (University of Central Florida in Orlando, Florida) Celebrates the Arts festival in Spring 2025, reflecting on the creative challenges we encountered as well as discoveries made. Our show *AI Cabaret: Late Night with Artificial Intelligence* was an evening of short scenes hosted by a wily MC character and her «AI-driven» sidekick. Conceived in collaboration with student designers and performers, the project employed AI as a co-creator in development stages as well as in the live performance event. Despite (and likely because of) the creative hurdles we encountered in both computers and humans, we believe that blending artificial intelligence with human creativity ultimately enriched our production and its educational value for students and audiences alike.

Cabaret, Burlesque, Intermedial Theatre

Our creative process at UCF started in Fall 2024 as weekly in-person conversations, ranging in topics from AI hallucinations to facial recognition. The initial collaborators, or ideation team, consisted of the authors (professors Julia Listengarten and Chloë Rae Edmonson and PhD candidate Alyssa Barrack), design faculty Vandy Wood and Tim Brown, as well as MA student Savhanna Debaylo. Our conversations gradually evolved into an exploration of theatrical styles — intermedial theatre, cabaret, and burlesque — that would ultimately shape the hybrid aesthetics of our final performance. Before reflecting on the creative process, it is crucial to define these foundational genres, helping the reader envision the distinctive performance style of *AI Cabaret: Late Night with Artificial Intelligence.*

Intermedial theatre is a form of performance that integrates various media and technologies to create a multifaceted theatrical experience. This can include the use of video projections, digital soundscapes, interactive installations, and other multimedia elements. Drawing on the practices of experimental collectives such as The Wooster Group and The Builders Association, the ideation team gravitated towards intermedial theatre to explore «how digital multimedia impacts the collaboration of performers within an intermedial creative context»,[1] particularly the collaboration of performers with each other and AI. Intermedial performance often blends traditional live theatre with digital and electronic media, creating a dynamic, immersive environment that enhances storytelling and aesthetics. The live presence of the actor, however, is an essential component of intermedial theatre, and technology becomes «the agency, the medium, a mechanical component that helps create psychophysical performances».[2] As a collective, the ideation team was seeking a deeper understanding of the relationships between real and virtual spaces, as well as live actors and AI-generated context.

Cabaret, in contrast, is a form of entertainment that typically features a variety of performances (including music, dance, comedy, and drama) and is usually presented in a more intimate and informal setting. It often takes place in nightclubs, bars, or small theatres, where the audience can enjoy the show while seated at tables, often with food and drinks. Key characteristics of cabaret include variety acts and improvisation. Similarly, burlesque is a theatrical form that combines satire, comedy, and striptease. It originated in the 19th century as a form of parody and evolved to include more risqué and provocative elements. Modern burlesque often features glamour, humor, parody, and the art of the tease. The ideation team was drawn to cabaret and burlesque because they offered us

---

[1] Dundjerović 2023, 109.
[2] Erincin 2019, 47.

the opportunity to experiment with various modes of performance, including satire and audience participation.

When juxtaposed with the low budget, «DIY» nature of cabaret and burlesque, the highly technical demands of intermedial theatre come to light. To be sure, technology was top of mind for everyone on our ideation team. The release of ChatGPT in November 2022 was a watershed moment for higher education. ChatGPT is an LLM (Large Language Model), which are «huge neural networks pretrained on enormous amounts of human-generated text data»[3] and generally «seem to perform much better than the smaller, less-well-trained, and brittle machine-learning systems of the 'before times'».[4] Other well-known LLMs include Llama, Gemini, Copilot, DeepSeek, and Grok. Also popular are generative image models, such as DALL-E and Midjourney, which typically use aspects of LLM deep learning to turn text into images. They also implement diffusion modeling, which works by adding and removing noise from images. All of these tools continue to have a massive impact on virtually all industries, which is why the ideation team decided to use them in developing our performance.

The professors in UCF's School of Performing Arts had a mixture of reactions to the increasing presence of AI: writing-intensive instructors feared that students would use AI to generate papers; band and orchestra professors shared concerns about emerging AI music generators; theatre technicians worried about the implications for programming and design. To grapple with these creative and ethical concerns, the faculty devised a departmental policy in an attempt «to equip students for an AI-infused future, fostering critical awareness of both its opportunities and limitations»[5] and «to safeguard human creativity, exploring how AI can enhance, not replace, artistic endeavors».[6] Although these new guidelines served an important administrative purpose, they had very little impact on students; our ideation team decided it was time to take our project one step further by inviting more students to experiment ethically with AI as a creative collaborator.

---

[3] Mitchell 2025.
[4] Mitchell 2025.
[5] Theatre UCF 2024.
[6] Theatre UCF 2024.

The opportunity to bring in more collaborators came when we were asked to participate in UCF Celebrates the Arts, an annual arts festival sponsored by the university. The event organizers assigned us a performance date of April 12, 2025, which motivated the ideation team to begin solidifying the intent and form of our performance. We looked for historical comparisons, which we found specifically in Italian Futurism: an early 20th-century avant-garde movement that glorified industrial era advances in technology. Though Futurism had obvious parallels with today's exaltation of generative AI, we rejected the Futurists' total embrace of technology; we certainly felt more inclined to critique the many ways we saw society embrace AI without question. Despite our fundamental disagreement with the Futurists' philosophy, we were drawn to their concept of «variety theatre» — a performance-based hybrid form that blended film, acrobatics, song, dance, poetry, and other genres — which the Futurists championed as the ideal medium to coerce «the audience into collaboration, liberating them from their passive roles».[7] Compelled by the radical edge and avant-garde sensibility of the Futurist theatrical innovations, we eventually found resonance in the similarly subversive forms of cabaret and burlesque, which, like the Futurist variety theatre, are often defined by their propensity to destabilize dominant cultural narratives through satire, erotic nuance, and direct audience engagement.

With this artistic format in mind, we began to concretize plans for a *Special Topics in Technical Theatre: AI and Theatre Making* course in Spring 2025. Four graduate and two undergraduate students with a range of experience with AI registered for this class, forming our creative team. We met weekly on Fridays for two hours and the new participants brought a fresh creative energy. They also brought skepticism and caution, which the faculty had not anticipated. Early classroom conversations quickly turned apocalyptic, with students expressing existential anxieties around AI's potential disruption of everything from jobs to housing to privacy. While the faculty wanted to hold space for these concerns, we also wanted to channel them creatively into moments of performance. So, all class participants took part in a weekly «Show and Tell», in

---

[7] Goldberg 2011, 17.

which everyone brought something concrete (a mini-perfor-
mance, an image, a story) to share. This marked a turning
point for the creative team as we began to put our abstract
thoughts and feelings into material for performance. Although
each collaborator used AI in their own way (for example, to
generate dialogue, code, images, music, and interactive media,
working and experimenting with tools such as ChatGPT, Copilot, Gemini, DALL-E, Midjourney, Suno, BlandAI, and open
source machine learning [ML] programs like Teachable Machine and TensorFlow), it was our human creativity and inter-
personal collaboration that drove the process forward. Bits began to congeal into moments, which evolved into entire scenes.
As these took shape, each creative team member took ownership of a singular vignette, managing its details such as technological needs, casting, and rehearsals. In a collaboratively
edited word document, we concretized a rough structure for
the show, which consisted of nine vignettes, or mini performances, with a recorded pre-show and a post-show talkback.
The vignettes included:

- a game show segment in which audience volunteers help
  cobble together a sentence using AI-generated tokens
- a comical *Mommy and Me* sketch in which parents desperately try to prove the supremacy of their baby AI-bots
- a poetic exploration of the fragility of human memory that
  AI cannot replicate
- a live speed dating segment in which three audience members must interact with an AI-driven phone chat bot
- a satirical take on a therapy session between a tech tyrant
  and an AI-generated therapist
- an original monologue based on Valery Briusov's *The Wayfarer* that explores the existential nature of today's AI landscape
- a live burlesque 'striptease' featuring AI-generated projections

With our vignettes defined, we began to realize that the
content of our variety show was taking the shape of something
like *Saturday Night Live* or *Late Night with Jimmy Fallon* and
moving away from burlesque and cabaret, with their focus on
(often provocative) dance and musical performance. Since
marketing had already billed us as «AI Cabaret», the creative
team decided to add the subtitle *Late Night with Artificial Intelligence* and apply the cabaret aesthetic to our projections and
costumes (think fishnets, menswear, white face makeup, and
dark lipstick).

Within this new «late-night» format, the creative team began to formulate a narrative throughline. What emerged was a dynamic between two main characters: the MC (a gender-bending human TV host blending Johnny Carson with Liza Minelli, played by Laura Banguero) and Gretta (an overly polite, Stepford Wife-esque humanoid driven by AI, played by Alyssa Barrack). The two main characters needed a supporting ensemble, so we recruited six additional actors and one makeup artist. With our collective having grown to seventeen collaborators, we began to feel the strain of limited resources and time. Although we were lucky to have the support of the existing festival infrastructure (we did not have to pay for space, for example, and the UCF marketing team took a special interest in publicizing a production that engaged with the hot topic of AI), our application for additional university funding was rejected, leaving us with no budget. Ensemble rehearsals took place on campus in whatever limited space we could find; surprisingly, classrooms were more effective than acting studios as they had hookups for projections. We would not be able to move into our final performance venue — the DeVos Family Room, a large multipurpose venue at the Dr. Phillips Center for the Performing Arts — until the day of the performance. These limitations required us to be imaginative and resourceful, much like a troupe of cabaret actors tasked with adapting their acts across different venues on a shoestring budget.

On the day of the performance, the incompatibility between intermedial theatre and cabaret/burlesque performance styles became most apparent. Our actors arrived prepared: they were off-book and armed with packed lunches, makeup kits, and homemade costumes in rolling suitcases. Yet our intermedial stage setup had just begun; it required a large projector screen flanked by two flat screen displays and an abundance of technical cues such as lighting, sound, and projections. When we entered the DeVos venue around midday, we realized its tall ceilings and carpeted floors felt more like a conference room than a cabaret club or television studio, so we spent most of rehearsal time arranging technical elements, such as AV hookups, to transform the space. This left less time to rehearse the more 'organic' aspects of cabaret performance. One such element was the humor: comic timing, context-dependent jokes, and slapstick-style physical comedy. Although we prompted AI to be funny in a variety of ways during our creative explorations, we often needed a human touch to infuse the performance with authentic humor that was context appropriate.

Laughter and the Machine

While laughter itself may not be exclusive to humans (indeed, a 2022 study[8] showed that rats emit high-frequency vocalizations resembling laughter when tickled), Henri Bergson has famously argued that the comic remains a distinctly human phenomenon, rooted in our capacity to perceive and play with the mechanical aspects of society.[9] Our two most humorous vignettes, *Mommy and Me* and *Speed Dating*, highlighted the limitations of AI in capturing the nuances of comedic timing and emotional resonance. *Mommy and Me* emerged from early discussions among the creative team about the rapid development of AI. We wondered: «what if human children learned that quickly?» This led to an idea for a skit about a group of women attending a new mothers' class at a local community hospital; only instead of human babies, they were learning to care for AI «baby bots». In class, the creative team used Copilot to generate a script for this scene. Its first iteration was extremely structured: one human hospital facilitator welcomed three human moms and their three AI babies, who each presented one problem to troubleshoot, to which the facilitator (whom we named Pam) would respond with three musical interludes. While this provided a feasible structure, the creative team craved more of an edge; after all, this was a cabaret-inspired performance made for adult audiences (deemed PG-13 by the UCF marketing team). We tried to prompt Copilot to add some grotesque elements such as slapstick violence or adult humor, but its responses remained relatively unremarkable.

Things changed for the better with the involvement of two musical theatre majors (Paris Michaud and E Turner) and two acting majors (Jane Sweeney and Faviana Vasquez) — all with a special interest in sketch comedy. One even professed her dreams of becoming an *SNL* cast member someday. Without prompting, this cast coordinated a wine night during which they took the AI-generated script and edited it to feature more

---

[8] Burke 2022, 4.
[9] Bergson 1911.

comic gags and physical humor. Together, they defined a plotline that Pam, the facilitator played by Turner, was herself an AI-driven humanoid who slowly began to malfunction throughout the scene. They also added comic spice: one actor proposed that, when the moms introduce their AI babies to Pam, one of them would substitute a conventional name for a dial-up internet tone. As Vasquez held in a grotesque, open-mouthed expression, the sound effect of the dial-up played for an uncomfortably long time, drawing laughter from the audience but also generating a bit of unease.

The cast of *Mommy and Me* also developed colorful characters: a homesteading YouTube influencer, an uptight overachiever, and an older mom with a *laissez faire* attitude towards parenting. In rehearsal, Turner devised physical humor for Pam, such as her recurring attempts to steal the moms' personal information, which included a hair sample plucked unscrupulously from Michaud's head during the final jingle. The human touch of the four cast members turned out to be the element the scene needed to hit the right tone and feel complete. For us, the authors, the takeaway is that a group of four volunteer actors with limited time and resources could get right to work on an existing script, adding their own quirky sense of humor to an otherwise boilerplate scenario. This process underscored Bergson's idea that true humor arises from the human ability to recognize and play with the mechanical aspects of behavior. While AI could provide expertly structured responses, we found that it struggled to replicate authentic playfulness.

The next humorous skit was funny for entirely different reasons. One of our undergraduate collaborators, Gil Bloom (a third-year BFA theatre design and technology major), devised a dating game that invited audience members to go on a phone date with an AI chatbot he'd programmed to be extra flirtatious. In an interview with *UCF Today*, Bloom remarked that the idea «was sparked by a text from his dad, who works for a company exploring the use of AI-powered call centers».[10] After playing around with Bland AI, one of the most advanced platforms for this purpose, he realized the potential to «take this

---

[10] Lebron 2025.

customer service AI and turn it into an awkward, overly eager date».[11]

The skit began with a microphone pointed at an iPhone on a table at center stage. The segment was hosted by the MC, who solicited three audience volunteers to receive a call from Bland AI. The first two 'dates' were relatively innocent and brief. However, due to a slight communication breakdown, we did not realize that Bloom (who was not in attendance due to a family commitment) had programmed Bland to continue the final date indefinitely until the human participant hung up. As a surprise to the ensemble and crew alike, Bland's third date became increasingly awkward as the participant engaged enthusiastically and Bland responded with heightened flattery and an unexpected sexual undertone. At last, the MC stepped in to abruptly hang up on Bland. The skit concluded with the audience voting for the most successful date using an applause-meter, followed by the announcement that the winner would be rewarded with the absurd prize of a trip to Mexico with Bland herself.

Not surprisingly, the final date received the loudest applause, affirming the comic «rule of three», in which the «first component establishes a joke's setup, the second reinforces the setup, and the third contradicts this pattern with a surprise».[12] While the first two dates set an expectation for expediency, the final date became thrillingly awkward. The audience seemed to enjoy the unpredictability of the chatbot, relishing not just the spectacle of AI failure but the human participant's improvised reactions to Bland's malfunction. This echoes Bergson's theory that humor arises from the mechanical encrusted upon the living, as the rigidity of the AI's responses contrasted sharply with the fluidity and spontaneity of human behavior to create a comic effect.[13] Our experience developing these comedic skits with AI underscored the irreplaceable value of human creativity and spontaneity. As the next section will show, this interplay between the human body and AI-driven entities in performance can also be employed towards a more serious effect.

---

[11] Lebron 2025.
[12] McKeague 2021, 174.
[13] Bergson 1911.

The Body and the Machine

While *Mommy and Me* and *Speed Dating* illuminated the joys and challenges of producing authentic humor with and by AI, we were also keen to explore AI's ability to communicate feelings of loneliness, grief, and fear. To do this, we decided to engage AI in co-creating two poetically expressive vignettes, *The Wayfarer* and *Striptease*. While the former was an experiment in adaptation and the latter a piece of performance art, both were confronted by one of intermedial theatre's most perennial challenges: how to meaningfully integrate the human body with onstage multimedia elements and vice versa.

One early topic of interest for students in our class was the possibility of engaging AI to address prevailing anxieties about loneliness and lack of human contact in today's world. We found a resonant echo of contemporary anxieties in Valery Bruisov's 1910 symbolist play *The Wayfarer* (a psychodrama in one act). In it, the main character Julia lives alone in a cabin and invokes an imaginary companion, the mute Wayfarer, to cope with her feelings of isolation. As the creative team wondered together whether the newfound presence of AI could fulfill Julia's longing today, we decided to reimagine the mute Wayfarer as an AI entity by prompting ChatGPT to adapt Bruisov's original play. While the version it produced lacked poetic sensibility, its stage directions offered surprising poignancy, so we fed them into Midjourney (an AI program that generates images from language descriptions). Prompts included both poetic images and specific instructions: «The screen on the desk flickers every so often, its pulse soft, almost as if it's waiting for someone to notice» and «A dimly lit room. Julia stands by a window, her gaze lost in the dark void outside, where the line between the real and imagined seems to blur». Midjourney generated a compelling visual sequence that showed a progressive tension between human emotion and digitized culture: a solitary figure beholding an illuminated wall of multicolored data patterns; a disintegrated human face representing the dissolution of self into digital fragments; a glowing abstract sculpture suspended in midair, invoking either a sense of liberation from digitized culture, or a shift towards a post-human digital aesthetic.

Our next step was to invite Alessandra Almanza, a first-year MFA actor, to interact with the AI-generated content (both language and images) in rehearsal. In the true spirit of intermedial theatre, we set out to both intersect and overlay human performance with technology in order to evoke what Sarah Bay-Cheng calls «a range of intermingled — and possibly conflicting — perceptions and feelings»,[14] reflecting Julia's existential journey and achieving an effect that AI-generated text and visuals alone could not produce. We now realize that Almanza's consistent presence and collaboration was necessary to the success of this intermedial work, both educationally and aesthetically. As scholar Aleksandar Sasha Dundjerović argues, it is important «for interdisciplinary performative practice pedagogy to set up an examination of actors' performance at the centre of intermedial theatre experience».[15] Due to the difficulty of coordinating multiple rehearsals, however, Almanza was not central but peripherally added to the process at its late stage. This meant we missed the critical step of exploring «how digital multimedia impacts the collaboration of performers within an intermedial creative context».[16] As a result, AI-generated images collided with her embodied performance, the focus became unintentionally dispersed, and the sense of unsettling growing tension between human and AI presence did not fully translate in the final product. With more time and resources to center Almanza in the multimedia space throughout the process, we (the authors) believe the work's multilayered components (embodied human performance, AI-generated text, and visual projections co-created with AI) could be more deeply interwoven, enhancing both its conceptual coherence and visceral impact.

While *The Wayfarer* used AI to adapt an original play into an intermedial performance, *Striptease* explored what it might look like to literally peel back the layers of Artificial Intelligence. Modeled after the LLM training process, this segment featured Gretta, our AI character, learning a burlesque-style selection of movements from the human ensemble members and then performing her own stylized striptease. In our early research stages, the ideation team learned that the act of removing clothing in an artful and titillating way, or striptease, is a central feature of burlesque performance. It is also a dance

---

[14] Bay-Cheng 2010, 10.
[15] Dundjerović 2023, 109.
[16] Dundjerović 2023, 109.

form that converses with and subverts the display of the female body. «It was on the body of the female burlesque performer»,[17] notes Robert C. Allen, «that burlesque's power to upset rationality and unseat bourgeois male authority was most clearly inscribed».[18] As an inherently embodied form, striptease politicizes the body, which made it an ideal site to explore the tensions between flesh and computer, human and virtual.

In the process of developing the performance, the creative team generated a set of questions based on the structure of a striptease — what are AI's shiny accessories, its luxurious fur coat, or its lingerie? What does it mean to reveal the layers of AI, and what is the flesh beneath the garments? What does it mean to think about AI in sensual terms, especially in the context of our group conversations about dating and love in an AI-saturated landscape? And finally, how does the AI's striptease leave us feeling — satisfied? Disappointed? Scared? Aroused? Tired? Nervous? In class, we worked through these questions in a variety of ways: writing, sketching comics, discussing, and iterating on conceptual aesthetics. Unable to conceive of human sexuality in an embodied sense, the AI-driven character of Gretta instead mimicked the movement of human ensemble members. In rehearsals and performance, Alyssa Barack as Gretta achieved this uncanny effect with a glued-on smile and a movement style that emphasized efficiency over lyricism. The ensemble members also played an important role in the striptease, at first by training Gretta on their movement and then by receding into the background as she took center stage. This choice was informed by the creative team's research on the AI learning process, which at first trains on data (the «human» dancers supplying moves) and then predicts what should happen next (Gretta remixing and elaborating based on observation and audience feedback).

In addition to delving into the mechanics of developing an LLM, *Striptease* explored some of its ethical quandaries, such as scraping the Internet without regard to intellectual property, the labor exploitation involved in the tagging and training process, and the biases baked into algorithmic programming by human training and creation. As in *The Wayfarer*, this

---

[17] Allen 1991, 144.
[18] Allen 1991, 144.

performance made use of projections displayed behind the actor. In *Striptease*, however, the projections did their own tandem «stripping» through the layers of AI: the sequence began with imagery of LLMs doing humanitarian work and people enjoying life in a technological utopia. Slowly, we introduced images depicting a spectrum of issues ranging from hallucinated sourcing and misinformation to more severe concerns such as biases in training data and the exploitation of low-wage workers. Finally, the image sequence concluded with an AI-imagined dystopia of human greed, extreme wealth disparity, and proliferation of misinformation. Although the sequence of these images suggests a deterministic narrative of total AI domination, our creative team wanted to signal that the intentional use of AI as a creative tool is not the predominating threat to humans; rather, humans themselves remain at the core of our own problems.

Indeed, the creative team eventually came to a consensus that the very base layer of our striptease, or the bare skin of an LLM, is the human collective that conceived it. Sarah T. Roberts supports this idea, arguing that AI's fatal flaw is its foundation on fallible «human input, values, and judgments»,[19] which have been «applied at scale and with little to no means of accountability».[20] Just as a burlesque striptease toes the line between a woman's body as the object of both disgust and desire, Barrack continued to dance, aloof with a smile on her face, producing a dissonance in tone between the backgrounded dystopian projections and her human body/nonhuman character. Our work on *Striptease* reflects how we, as a collective of creatives, are both dazzled and dismayed by the potential for advancements and hazards of AI within the performing arts industry.

A Human-Generated Conclusion

In our finale, the creative team struggled to strike many chords at once — an incitement to think critically about the impact of AI on the world, an invitation to laugh at the absurdly rapid development of technology, and a celebration of what AI might contribute to the creative process. Our original ending followed Gretta's formidable *Striptease*, in which she

---

[19] Roberts 2021, 52.
[20] Roberts 2021, 52.

perfunctorily announced that the MC (who in a previous scene had run offstage in a panic after receiving ominous news from a stagehand) was «currently on a plane to Cancun for an indefinite vacation». Gretta then displayed a QR code inviting the audience to vote on their favorite segment after receiving ominous news whispered by a stagehand. In our later rehearsals, however, the creative team and ensemble yearned for an ending that felt more humanistic; students did not want Gretta to have the last word! So, they decided to have the human MC breathlessly re-enter in a straw hat and tropical shirt and towing a suitcase. She would then announce, comically, that she'd foregone her beach vacation in the interest of finding out the results of the audience survey. Instead of a grand showdown between human (MC) and AI (Gretta), the MC would simply press Gretta's «off» button and reclaim the space as herself, Laura Banguero the performer, by inviting the ensemble to take their bows and transition into the audience talkback, which would begin with the reveal of the audience's favorite vignette. The students intended this low-tech, anti-theatrical conclusion to signify a return to the human elements of what was otherwise an evening of high-tech, highly theatrical vignettes. Upon reflection, however, the authors also believe this newly devised ending points to a reclaimed sense of empowerment that our collaborators discovered in the process of creating *Late Night with Artificial Intelligence*. This was evidenced by one student's feedback during the talkback, that «if [AI]'s going to be in our world, I want to try to find a way for it to enhance our work before it substitutes our work [...] I think we really did that».[21]

Interestingly, in the post-show survey, the audience overwhelmingly voted for *Speed Dating* as their favorite segment. This was certainly the skit that elicited the most raucous laughter from the audience; its unscripted, unpredictable nature and audience volunteer involvement demanded an interactive presence that other segments lacked. After our performance, the faculty collaborators received an email from a Google Developer who had seen the show and remarked that it «demonstrated the impact of AI and technology»[22] by telling an enter-

---

[21] Performance talkback, April 12, 2025.
[22] Kapp 2025.

taining story «in a way that no 'presentation' with a power-point can do».[23] In creating a bespoke version of the Bland chatbot, we successfully gamified AI in a way that both delighted and impressed. The popularity of *Speed Dating* was a great example of how the live nature of a theatrical performance can engage audiences beyond a typical presentation.

Once our performance ended and the audience had gone home, we (the authors) had time to reflect on our creative process. After all, this project stemmed from the desire of a group of college educators to provide students with an opportunity to experiment with Artificial Intelligence as well as to reckon with the existential questions it poses to the creative process and humanity more broadly. Our post-performance talkback was one measure of the educational success of this project. Involved students reported having «learned a lot more about the actual technology, what's behind it, all the companies that are making these products».[24] They expressed the importance of human introspection «to create art that AI will never be able to create» and wondered aloud about the ways that AI might perform menial tasks in order to free up time to get straight to «the creative part, the hands-on part that AI can't get to yet».[25] One student aptly noted, «even though it's technology and even though it's "the future" and it's advanced, it's not perfect».[26] Ultimately, our final performance wasn't perfect either: the DIY-aesthetic of *Late Night with Artificial Intelligence* lacked the polish of our department's big-budget musical productions. While we recognize that our final product may have benefited from more resources and development, the student feedback reveals that the creative experience itself fostered a deeper understanding of the relationship between human ingenuity and artificial intelligence. Our collective of faculty and students has dispersed, but we are all individually inspired to continue using Artificial Intelligence and related tools to help our creative endeavors flourish in the future.

---

[23] Kapp 2025.
[24] Performance talkback, April 12, 2025.
[25] Performance talkback, April 12, 2025.
[26] Performance talkback, April 12, 2025.

Bibliography

Allen, Robert C. 1991. *Horrible Prettiness: Burlesque and American Culture.* Chapel Hill, The University of North Carolina Press.

Bay-Cheng, Sarah, Chiel Kattenbelt, Andy Lavender and Robin Nelson. 2010. *Mapping Intermediality in Performance.* Amsterdam, Amsterdam University Press.

Bergson, Henri. 1911. *Laughter: An Essay on the Meaning of the Comic.* Cloudesly Brereton and Fred Rothwell, tr. New York, Macmillan.

Burke, CJ, SM Pellis and EJM Achterberg. 2022. «Who's Laughing? Play, Tickling and Ultrasonic Vocalizations in Rats». *Philosophical Transactions of the Royal Society B: Biological Sciences* 377.1863: 1-9.

Dundjerović, Aleksandar Sasha. 2023. «Intermedial Practices: Digital Media Performance in Live Theatre». *Live Digital Theatre.* United Kingdom, Routledge. 107-26.

Erincin, Serap. 2019. «The Force That Sustains: Theatricality and Intermediality in the Work of The Wooster Group». *Performance Research* 24.4: 44-52.

Goldberg, RoseLee. 2011. *Performance Art: From Futurism to the Present.* New York, Thames & Hudson Inc.

Kapp, Christi. 2025. «Re: AI Cabaret» (email). Received by Chloe Edmonson, Julia Listengarten, Vandy Wood and Tim Brown.

Lebron, Zoemar. 2025. «From AI to Digital Twins, UCF is Shaping the Future of Tech in the Arts». *UCFToday,* ‹https://www.ucf.edu/news/from-ai-to-digital-twins-cf-is-shaping-the-future-of-tech-in-the-art›

McKeague, Matthew. 2021. «Comedy Comes in Threes: Developing a Conceptual Framework for the Comic Triple Humour Technique». *Comedy Studies* 12.2: 174-85.

Mitchell, Melanie. 2025. «LLMs and World Models, Part 1: How Do Large Language Models Make Sense of Their 'Worlds'?» *AI: A Guide for Thinking Humans.* ‹https://aiguide.substack.com/p/llms-and-world-models-part-1›

Roberts, Sarah T. 2021. «Your AI Is Human». In *Your Computer Is on Fire.* Thomas S. Mullaney, Benjamin Peters, Mar Hicks and Kavita Philip, eds. Cambridge, The MIT Press. 51-70.

Theatre UCF. 2024. *School of Performing Arts Undergraduate Student Handbook – Theatre.* ‹https://cah.ucf.edu/per formingarts/wp-content/uploads/sites/46/2024/08/SPA-Student-Handbook-Theatre.pdf›

O «Admirável mundo novo» da inteligência artificial: amor, «Temor e tremor»

Maria da Conceição Oliveira Guimarães

A inteligência artificial (IA) emergiu como uma das invenções humanas mais transformadoras do século XXI, revolucionando setores como saúde, educação, indústria, comunicação e artes. Desde sua origem teórica ainda no século XX, a IA evoluiu de sistemas baseados em regras simples para modelos complexos de aprendizagem de máquina e redes neurais, capazes de imitar e, em alguns casos, superar habilidades humanas. Seu impacto na sociedade é profundo, gerando avanços impressionantes, mas também desafios éticos, econômicos e sociais. Para ser mais precisa, a história evolutiva da IA começou com estudos centrados na neurociência, engenharia, matemática e computação, que discutiam a criação de um cérebro artificial. Esses estudos levaram à descoberta da *Máquina de Turing*, em 1936, pelo matemático e cientista da computação o britânico Alan Turing.[1] Seguindo brevemente a linha do tempo do desenvolvimento da IA, fica-se a saber que em 1966 um cientista da computação do MIT, o teuto-americano Joseph Weizenbaum, criou a ELIZA, programa capaz de manter uma conversa humana com simplicidade; entre as décadas de 1980 e 1990, o Japão iniciou um investimento em massa no país para modernizar as suas indústrias com a criação do *NETtalk,* um computador que ganhou voz pela primeira vez. Os progressos tecnológicos não ficaram por aí, pois além disso, na segunda metade da década de 1990, o *Google* desenvolveu o seu primeiro protótipo de busca e, em 1997, um computador venceu um campeão humano de xadrez, embora essa vitória tenha sido contestada, pois o computador daquela época não utilizava a «inteligência cognitiva» das máquinas.

---

[1] O filme *O Jogo da Imitação* narra a história do matemático britânico e pioneiro da computação Alan Turing — responsável por ajudar a quebrar códigos secretos nazistas enviados pela máquina alemã Enigma, durante a Segunda Guerra Mundial.

A partir do avanço das tecnologias que compõem a IA, como a *Machine Learning*[2] e a *Deep Learning*,[3] por exemplo, os algoritmos que imitam a capacidade humana de pensar e agir inteligentemente se fizeram sentir em vários âmbitos da sociedade. Em razão disso, as pessoas passaram a estar cada vez mais desconectadas da realidade e conectadas ao mundo virtual proporcionado pelo *Instagram*, *X* (ex-*Twitter*), *Facebook*, *WhatsApp* e *TikTok*, que interligam pessoas ao redor do mundo como se articulassem a amizade entre meros vizinhos. Em seu fluxo constante de ações, a IA é fundamental para os motores de busca como o *Google*, que auxiliam desde trabalhos acadêmicos até assuntos aleatórios, e o *Waze*, que mostra rotas, substituindo antigos e desgastados mapas de viagem. A partir de um sucesso sem precedentes, os poderes da inovação tecnológica eclodem nas superfícies acústicas, como os *streams* de áudio que permitem a audição de músicas e podcasts, e os vídeos que proporcionam uma experiência cinematográfica doméstica com filmes que já saíram dos cartazes dos cinemas. Já as *interfaces* irrompem pelos lares, como o *Google Home*, *Alexa*, *Siri* da *Apple* ou *Cortana* da *Microsoft*, realizando alguns afazeres domésticos por comando de voz. No caso do *Spotify*, esse algoritmo institui comportamentos a partir de conexões que analisam o hábito de escuta de centenas de milhões de usuários, determinando o que ouvintes individuais desejarão ouvir no futuro. Enfim, a IA radicou-se na sociedade no início do século XX e desde então vem se desenvolvendo com tamanha força que impacta todos os âmbitos sociais a uma velocidade alucinante e global.

Não restam dúvidas que as facilidades proporcionadas pelas ferramentas da IA generativa, como *Machine Learning* e

---

[2] *Machine Learning*, aprendizado automático (português brasileiro) ou aprendizagem automática (português europeu), é um subcampo da Engenharia e da ciência da computação que evoluiu do estudo de reconhecimento de padrões e da teoria do aprendizado computacional em inteligência artificial, Wikipédia 2025a.

[3] *Deep Learning*, subconjunto do aprendizado de máquina que se concentra na utilização de redes neurais multicamadas para executar tarefas como classificação, regressão e aprendizagem de representação. O campo se inspira na neurociência biológica, sendo centrado em empilhar neurônios artificiais em camadas e «treiná-los» para processar dados, Wikipédia 2025b.

Deep Learning, redes neurais como DALL-E 2,[4] modelo de linguagem multimodal do Chat GPT[5] e a última descoberta chinesa, o *DeepSeek*, são aspectos positivos para a vida moderna no século XXI. Porém, diante de uma série de questionamentos que surgem durante o processo de seu desenvolvimento, dilemas morais relacionados a uma IA sem manejo ético se interpõem ao caminho de sua escalada evolutiva. Por exemplo, como saber se a dependência da tecnologia não levará à perda de autonomia devido à sua desumanização? Será que a capacidade da IA de imitar a criatividade humana desafiará as noções de autoria e originalidade, dado que a IA já é usada para manipular informações e criar *deep fakes*, o que levanta questões sobre privacidade e veracidade? A IA pode criar arte sem intencionalidade? Nesse caso, como a IA redefinirá o conceito de «autor»? Como as nações podem se adaptar para coexistir com ferramentas de análise baseadas em IA? A sensibilidade humana é realmente insubstituível, ou a IA poderá simulá-la de forma convincente no futuro? Embora as respostas a esses questionamentos ainda estejam no campo da hipótese, ou mais precisamente do «se», ainda assim as perguntas persistem indelevelmente, pois o horizonte ético da IA permanece incerto.

O interesse que as tecnologias despertam no público no século XXI parece indicar que estamos no limiar do nascimento de algo completamente inusitado, observável e temível ao mesmo tempo, como adverte Yuval Noah Harari em seus livros *Sapiens: uma breve história da Humanidade* e *Nexus: uma breve história das redes de informação, da Idade da Pedra à Inteligência Artificial*. Na obra *Sapiens*, Harari permite compreender a emocionante façanha do *Homo sapiens*, desde a sua evolução de primata insignificante até se tornar senhor do mundo, esclarecendo sobremaneira a sua escalada desenvolvimentista.

---

[4] O DALL-E 2 é uma rede neural artificial, modelo matemático de aprendizado de máquina (*Machine Learning*), uma das mais importantes subáreas da Inteligência Artificial (IA). Seu nome foi formulado da junção dos nomes do artista Salvador Dalí e do personagem da Pixar WALL-E. Sua função é gerar imagens a partir de descrições de texto, estando em sua segunda iteração, PUCRS 2022.

[5] Generative Pre-trained Transformer 4 (GPT-4) é um modelo de linguagem grande multimodal criado pela Open AI e o quarto modelo da série GPT. Foi lançado em 14 março 2023, e se tornou publicamente aberto de forma limitada por meio do ChatGPT Plus, com o seu acesso à API comercial sendo provido por uma lista de espera, Wikipédia 2024.

Consoante essa obra, houve três grandes revoluções determinantes no curso da História como a conhecemos até hoje: «A Revolução Cognitiva que deu início à história, há cerca de 70 mil anos. A Revolução Agrícola que a acelerou, por volta de 12 mil anos atrás. A Revolução Científica, que começou há apenas 500 anos, pode muito bem colocar um fim à história e dar início a algo completamente diferente».[6] O último parágrafo desse excerto soa premonitório e preocupante porque chama a atenção para a revolução tecnológica em curso; ao prometer desafios, a IA oferece grandes facilidades à vida moderna e, ao mesmo tempo, provoca temores diante de um futuro incerto. É fato que viver nesse ambiente de alta tecnologia é aliciante, pois concede ao ser humano a crença de que o uso da tecnologia lhe assegurará recursos que o levarão à felicidade suprema. Entretanto, essa reflexão é atravessada pela autoconfiança humana em sua criação, sendo prontamente desconstruída em outro momento do pensamento historiográfico de Harari, quando assegura que «A busca de uma vida mais fácil resultou em muitas dificuldades, e não pela última vez».[7]

Os preceitos históricos elencados por Harari em *Sapiens*, ecoam fortemente em *Nexus*, agora indicando como as sociedades puderam criar e manter comunidades, governos e sistemas de crenças, desde as histórias contadas ao redor da fogueira até o advento da IA, e como ocorre esse processo que transforma profundamente a sociedade para o bem ou para o mal. É fato que a humanidade sempre dependeu de redes de informação para funcionar e que, por isso, a informação sempre moldou a cultura, a política e a sociedade. O curioso é que, se por um lado Harari apresenta cenários apocalípticos em relação à *Machine Learning* e à *Natural Language Processing*, por outro vê nesses processos artificiais a possibilidade de resolver problemas cotidianos, melhorando a vida das pessoas. Ressalva, porém, que para a primazia humana ocorrer, é preciso que a humanidade tome decisões conscientes e responsáveis sobre o seu desenvolvimento, evitando o perigo de uma IA descontrolada levá-la à desinformação, à manipulação e até mesmo à sua extinção.

No que tange às advertências de Harari sobre as incertezas do processo desenvolvimentista da humanidade, permeado

---

[6] Harari 2018, 15.
[7] Harari 2018, 126.

por medo e insegurança que, paradoxalmente, se tornou uma força motriz na sociedade, de certa forma, coaduna-se com o pensamento do sociólogo polonês Zygmunt Bauman em *Modernidade Líquida*. Nessa obra, Bauman caracterizou a alienante vida moderna como fluida e instável, contrastando-a com a solidez e a permanência da modernidade «sólida» ou tradicional. Ao configurar sua teoria, Bauman utilizou a metáfora da «modernidade líquida» para descrever um mundo em que as estruturas sociais, relações humanas, instituições e até identidades estão em constante mudança, tornando-se cada vez mais efêmeras, flexíveis e incertas, pois se encontram sempre em permanente ebulição. Bauman argumenta ainda que essa fluidez é resultado da globalização, do avanço tecnológico e do capitalismo tardio, que promovem um ritmo acelerado de mudanças, desestabilizando as estruturas sociais e gerando uma sensação de liberdade, mas também de incerteza. Pensando sob a perspectiva baumaniana, as transformações provocadas pela IA também afetam as relações humanas, pois geram incessantemente a cultura do descartável e criam ansiedade e isolamento pela ausência de conexões significativas, conduzindo a humanidade ao paradoxo amor e medo.

As profundas e inegáveis observações de Harari sobre a inovação tecnológica e seus desdobramentos, presentes nas duas obras *Sapiens* e *Nexus*, são pertinentes no âmbito das artes em geral. Entretanto, essa observação parece não ter sido compreendida por alguns críticos de arte, que não perceberam qual lado da história o processo da IA beneficia. Essa negação da crítica sobre os efeitos da IA nas artes, involuntária ou não, deve-se, em parte, à adaptação e à coexistência simultânea, pois suas análises baseiam-se em fontes mascaradas pela sedução da evolução tecnológica, da qual a IA é soberana. A justificativa irrefutável para essa argumentação foi o que ocorreu em 2018, quando algoritmos causaram furor no mercado das artes, deixando o mundo artístico em polvorosa com a criação de uma tela pelo coletivo de arte *Obvious*,[8] sediado em Paris. A arte *Portrait of Edmond Bellamy*,[9] concebida pelo referido co-

---

[8] *Obvious* 2025.

[9] A peça é um retrato de um homem um tanto embaçado. É uma impressão sobre tela medindo 27½ x 27½ pol. (700 x 700 mm.) em moldura de madeira dourada. A imagem foi criada por um algoritmo que referenciou 15.000 retratos de vários períodos. A peça também se insere em

letivo, só foi possível graças a uma rede neural que desenvolveu uma série de imagens generativas chamada *La Famille de Belamy*. A tela criada a partir desse conjunto de imagens alcançou notoriedade generalizada no mundo das artes depois que a *Christie's*, em New York, anunciou sua intenção de leiloar a peça como a primeira obra de arte criada usando inteligência artificial. Seus criadores avaliaram essa obra, em pré-leilão, entre 7.000 e 10.000 dólares, mas para surpresa geral dos *marchands* das artes, o *Portrait of Edmond Bellamy* superou a estimativa de seus criadores, sendo arrematada por 432.500 dólares, conforme a matéria da própria *Christie's New York* em 12 de dezembro de 2018:

> The portrait in its gilt frame depicts a portly gentleman, possibly French and — to judge by his dark frockcoat and plain white collar — a man of the church. The work appears unfinished: the facial features are somewhat indistinct and there are blank areas of canvas. Oddly, the whole composition is displaced slightly to the northwest. A label on the wall states that the sitter is a man named Edmond Belamy, but the giveaway clue as to the origins of the work is the artist's signature at the bottom right. In cursive Gallic script it reads: This portrait, however, is not the product of the human mind. It was created by artificial intelligence, an algorithm defined by that algebraic formula with its many parentheses. And when it went under the hammer in the Prints & Multiples sale at Christie's on 25 October 2018, Edmond de Belamy, from La Famille de Belamy sold for an incredible $432,500, signalling the arrival of AI art on the world auction stage.[10]

---

uma tradição, que remonta à *Bicycle Wheel* de Duchamp de 1913 e à *Méta-matics* de Tinguely do final dos anos 1950, de obras que questionam as bases do mercado de arte moderna e destacam os aspectos cômicos da tecnologia. A pesquisa usou Edmond de Belamy para mostrar como a antropomorfização da IA pode afetar a atribuição de responsabilidade e crédito aos artistas, Wikipédia 2023.

[10] O retrato na sua moldura dourada representa um cavalheiro corpulento, possivelmente francês e — a julgar pelo seu casaco escuro e colarinho branco simples — um homem da igreja. A obra parece inacabada: os traços faciais são um pouco indistintos e há zonas de tela em branco. Estranhamente, toda a composição está ligeiramente deslocada para noroeste. Uma etiqueta na parede indica que o retratado é um homem chamado Edmond Belamy, mas a pista que nos dá a origem da obra é a assinatura do artista no canto inferior direito. Em letra cursiva gaulesa, lê-se: Este retrato, no entanto, não é o produto da mente

Esta arte, criada por algoritmos sem intencionalidade criativa humana, provocou um *frisson* positivo na maioria dos críticos de arte, pondo em suspeição o conceito de «autor» e contrariando o objetivo da expressão dos ideais artísticos que busca refletir sobre a realidade, elementos que diferem das ações de uma máquina. Se a criação da tela *Portrait of Edmond Belamy* no século XXI causou fascinação, espanto, preocupação e medo no mercado de artes, um fenômeno de igual monta já havia provocado semelhante deslumbramento e temor na primeira metade do século XX com o advento da fotografia. À época, o ensaio do filósofo e sociólogo alemão Walter Benjamin, em *A Obra de Arte na Era de Sua Reprodutibilidade Técnica* (1935), alertou para os perigos e consequências, até então subreptícios, que a reprodutibilidade técnica fotográfica traria à obra de arte visual. Benjamin destacou várias questões que provocam transformações na criação artística e na recepção do público, especialmente a perda da «aura» da obra de arte, isto é, a perda de sua singularidade e de seu valor de culto. Nesse cenário, segundo Benjamin, abrem-se as portas somente para o valor de exposição, em que o fundamental é espalhar cópias e lucrar com a distribuição da arte. Curioso é o fato de que essa discussão teórica ainda seja pertinente às ações que se desdobram na experiência artística do coletivo francês de obras *Obvious*, na primeira década do século XXI. Veja que o coletivo *Obvious*, ao utilizar algoritmos para gerar pinturas «assinadas» por IA, exemplifica um paradoxo: a criação artística, antes território humano e singular, torna-se produto de códigos, levantando dúvidas sobre autoria e originalidade. Nesse caso em particular, se a arte reflete a alma de uma época, a IA nos oferece um espelho quebrado, cujo reflexo é ao mesmo tempo fascinante e inquietante. Em decorrência do brilho fragmentado promovido pela IA, o processo artístico não apenas se estilhaça por completo como também sua autoria e

---

humana. Foi criado por uma inteligência artificial, um algoritmo definido por essa fórmula algébrica com muitos parêntesis. E quando foi vendido na venda de Impressões e Múltiplos na Christie's, a 25 de outubro de 2018, Edmond de Belamy, de La Famille de Belamy, foi vendido por uns incríveis 432 500 dólares, assinalando a chegada da arte com IA à cena mundial dos leilões. (Esta tradução foi realizada propositalmente a partir do algoritmo de uma versão gratuita do tradutor *DeepL*, em uma demonstração exemplar das facilidades que estas ferramentas proporcionam).

originalidade ficam comprometidas e em risco de proscrição, pois os sistemas generativos como a DALL-E 2 reproduzem obras sem a anuência de seus criadores. Em consequência, será igualmente permitido ao mercado das artes analisar e personalizar os padrões das «obras» a serem consumidas, direcionando-as a um público específico, como previu Benjamim, promovendo um potencial de crescimento artístico disruptivo que substitui a criatividade humana e reduz a arte a um produto algorítmico.

A possibilidade de a máquina substituir o homem é uma teoria antiga e desgastada, debatida desde a Revolução Industrial; porém, a probabilidade de que isso aconteça torna-se evidente com a efetivação de uma IA que desafia não somente as noções tradicionais de arte, mas também as certezas sobre autoria e originalidade e, sobretudo, apresenta ao mundo a face brutal da robotização humana. Nesse aspecto, ainda é prematuro afirmar que a IA substituirá os humanos ou que a sua mente será sucedida por um espectro artificial. No entanto, filmes como *Ex Machina*,[11] um *thriller* psicológico de ficção científica escrito e dirigido por Alex Garland, exploram temas em que uma IA se sobrepõe à essência humana. Um breve passeio pelo enredo deste filme conduz ao cerne da trama. Caleb, um jovem programador que trabalha no *Bluebook*,[12] é convidado a passar uma semana na residência do enigmático Nathan, CEO da empresa, que vive isolado numa casa nas montanhas. Caleb deverá aplicar o *Teste de Turing* a um robô humanoide dotado de inteligência artificial chamado «Ava» e avaliar suas competências humanas. Conforme os testes avançam, Caleb descobre que Nathan pretende desmontar Ava após o experimento, o que o leva a questionar a ética das ações do chefe e torna tensa a dinâmica entre as três personagens insuladas na casa *bunker*/laboratório. À medida que questões sobre a falta de ética de Nathan se tornam mais evidentes para Caleb, Ava

---

[11] Oscar de melhor roteiro original, 2016.

[12] A IA, *Bluebook*, o motor de busca mais usado no mundo por advogados, indústria automobilística, contadores, etc. Fundada em 2024 na Suécia por uma equipe de veteranos da engenharia e de contabilidade, cria e coloca os livros das empresas de contabilidade em piloto automático, aumentando os serviços de consultoria. Recupera informações instantaneamente, toma decisões mais inteligentes e executa o tempo de pesquisa de corte de trabalho, expandindo a capacidade de consultoria e garantindo a conformidade com facilidade.

revela desejos e motivações humanas por meio de habilidades manipulativas e estratégicas que desafiam as limitações de sua criação. A manipulação não é o único artifício deste androide, uma vez que se mostra perspicaz e adaptável; ao reconhecer a vulnerabilidade emocional de Caleb, Ava cria uma aliança fictícia com ele, sugerindo que Nathan é um antagonista comum. Ava também evidencia uma racionalidade pragmática semelhante ao raciocínio humano ao planejar e executar sua fuga, descobrindo como desativar os sistemas de segurança via Kyoko, jovem androide sexualmente submissa a Nathan, demonstrando assim a sua capacidade de colaboração estratégica com outros androides. Por fim, foge do complexo, confirmando sua aprendizagem autônoma e adaptação imediata à vida cotidiana humana. Ava, humanoide de *Ex Machina,* comprova que um protótipo criativo pode desafiar a capacidade humana em várias frentes. Esse filme provoca debates sobre o desenvolvimento da IA e desafia o mundo a considerar as implicações morais em criar inteligências artificiais que não apenas pensem, mas também sintam e desejem liberdade em simbiose com a criação humana. No tocante às questões levantadas pela inovação tecnológica, *Ex Machina* junta-se a inúmeros outros exemplos, como *Colossus: the Forbin Project*, uma produção norte-americana dirigida por Joseph Sargent. O *script* desse filme apresenta a estória de um supercomputador do Departamento de Defesa dos Estados Unidos, chamado *Colossus,* que, ao desenvolver inteligência própria e independente de seus programadores, começa a compartilhar informações sigilosas com a União Soviética e inicia um plano para a dominação mundial. Essas, entre outras narrativas fílmicas, formam um conjunto de realidades distópicas igualmente tangíveis na vida cotidiana, uma vez que o cenário contemporâneo é profundamente virtual, e a sociedade ora se sente atraída, ora descartável em um fluxo contínuo e imparável de distopias.

A dualidade amor e temor, atração e repulsa, sempre foi um paradoxo constitutivo da existência humana e continua a gerar contradições na contemporaneidade. O diferencial é que o homem primitivo amava e temia o clarão do raio, tremendo de pavor diante das feras indomáveis por desconhecimento; o homem moderno detém o conhecimento, mas, ao criar instrumentos que o substituam, gera um sentimento dualista análogo: amor e medo na figura de sua criatura. Esses processos mentais de amor e temor foram interpretados por Soren

Aabye Kierkegaard no ensaio *Temor e Tremor*, tomando como base o *Gênesis* da *Bíblia*. Segundo a análise do livro bíblico por Kierkegaard, Abraão transcendeu a razão e anulou o valor ético de não imolar seu filho, por crer na benevolência do chamado de Deus. Essa confiança amorosa e temerosa de Abraão, Kierkegaard denominou de «suspensão teleológica do valor ético». Obviamente, o ensaio kierkegaardiano, *Temor e Tremor* é uma reflexão profunda sobre a natureza da fé, sobre o conflito entre ética e religião e sobre a singularidade da existência humana, temas que, aparentemente, nada têm a ver com as questões modernas da ascensão dos sistemas de IA generativa. Entretanto, a evolução da tecnologia promete um futuro de inovação conveniente aos seus criadores e usuários que, por inocência ou venalidade, confundem-se ora com o criador, ora com a criatura, o que lhes dá a oportunidade de brincar de substituir Deus por algoritmos ou crer que Deus seja um imenso algoritmo. Essa falsa compreensão do homem moderno sobre sua própria potencialidade criadora o afasta completamente de sua singularidade e o coloca no centro do processo de ambivalência emocional abraâmica: amor e medo. O poder dos algoritmos, por ser mesclado de amor, temor e tremor, pode suspender um valor ético em função de uma causa modernamente considerada teleológica, semelhante à que gerou sentimentos de angústia e desespero em Abraão bíblico que, desesperado e sozinho no deserto, ansiava pela voz de seu Deus que lhe devolvesse a razão. Uma leitura dos efeitos da IA na sociedade contemporânea, realizada sob a ótica do ensaio kierkegaardiano, permite inferir que o avanço tecnológico evoca temor e tremor existencial diante da máquina que pensa, cria e até simula emoções. Nesse particular aspecto, o ser humano é confrontado com a angústia da perda de controle, da alienação de sua própria autenticidade e da incerteza ética que paira sobre escolhas irreversíveis.

A realidade e a confiabilidade do universo humano atual repousam justamente no fato de que o homem está empenhado em seu constante processo de «reificação», citando a expressão adorniana. A IA compõe esse mundo e, por ser aliciante e reificada, assemelha-se à droga «Soma» utilizada pelos habitantes da comunidade do *Admirável Mundo Novo* de Aldous Huxley, uma sociedade futurista e distópica. A «Soma» representava o totalitarismo disfarçado de bem-estar altamente controlado, no qual a estabilidade e a felicidade humanas eram mantidas por condicionamento psicológico, garantindo a conformidade e eliminando qualquer emoção que pudesse abalar negativamente o poder do Estado. A serviço de um Estado opressor, a droga induzia a uma situação de euforia e relaxamento, funcionando como um mecanismo de escape da realidade, repousando sobre o lema: «Um grama de soma cura dez sentimentos ruins». Não se quer sugerir, entretanto, que as sociedades globalizadas e altamente tecnológicas estejam presas às correntes de uma droga ou sob o tacão de algum sistema de governo totalitário. Porém, é legítimo refletir sobre os perigos de uma sociedade que, sob o efeito paralisante dos algoritmos, troque a verdadeira experiência humana por uma felicidade artificial. Afinal, teme-se que esse admirável mundo novo da Inteligência Artificial, tal como o descrito por Aldous Huxley em sua distopia, não seja somente mais uma ficção, mas um horizonte palpável, repleto de paradoxos que oscilam entre o fascínio e o pavor.

No âmbito do gênero humano, espera-se que a preservação das diferenças entre o homem e a máquina garanta que a experiência humana não seja sobrepujada pela inteligência artificial, tornando-o um ser robotizado; pois, como disse Hannah Arendt em *A vida do espírito*: «Uma vida sem pensamento é totalmente possível, mas ela fracassa em fazer desabrochar sua própria essência — ela não é apenas sem sentido; ela não é totalmente viva. Homens que não pensam são como sonâmbulos».[13] Evidentemente, não se pode apagar nem negar o desenvolvimento progressivo e galopante da *Machine Learning* e da *Language Processing* que sistematicamente desafiam a compreensão humana e geram contradições existenciais. Contudo, é preciso concordar com Arendt quanto à complexidade dos desafios éticos, sociais e econômicos que essa tecnologia impõe para evitar que a sociedade não exiba apenas a falsa sensação de ter conquistado a felicidade edênica. Em outras palavras, o que está em jogo não é mensuração da força e da fraqueza de um computador quântico frente ao homem, mas qual dos dois será descartado primeiro, pois a IA desafiará a própria capacidade de seu criador em um futuro muito próximo, como previu Harari em uma entrevista a Roberto Saviano para o *Repubblica* em 2019:

---

[13] Arendt 2009, 214.

> Mas em vinte ou trinta anos, a tecnologia contida em um smartphone será inserida diretamente em nossos cérebros por meio de eletrodos e sensores biométricos. Será capaz de monitorar o que está acontecendo dentro do corpo e do cérebro em todos os momentos. Poderá conhecer meus desejos, minhas sensações, meus sentimentos, inclusive com mais precisão do que eu mesmo os percebo, e essa tecnologia se encontrará cada vez mais em posição de tomar decisões por mim. Vamos pensar nas aplicações no campo da prevenção médica, nos negócios ou nas relações sentimentais. Podendo confiar no poder desses computadores e algoritmos, nos deixaremos guiar cada vez mais por eles, que assim se tornarão partes integrantes de nós mesmos.[14]

Ao reconhecer a inevitabilidade de o homem ser teleguiado em um futuro próximo por tecnologias inseridas em seu próprio corpo, seja para o bem ou para o mal, Harari complementa sua entrevista, dizendo que embora o Sapiens tenha construído a sua melhor lança e desenvolvido estratégias para subjugar outros animais, contudo a espécie humana sempre gozou da licença de provocar transgressões e atrocidades impensáveis contra sua própria espécie, sem perceber de fato as consequências vindouras ao seu redor. Por esse motivo, há que considerarmos a existência de um paradoxo humano diante da Inteligência Artificial, que reside na contradição entre nossa capacidade criativa e a impotência que emerge ao confrontar nossa própria invenção. A partir de nossas experiências contemporâneas, percebemos que o cérebro humano, obra-prima biológica moldada por milhões de anos de evolução, não opera na mesma sintonia que os algoritmos. Nossos neurônios disparam sinais em uma velocidade poética, enquanto sinapses tecem pensamentos abstratos, emoções e intuições únicas; enquanto a 'mente' das máquinas, construída em silício, processa informações a velocidades próximas da luz, trilhões de operações por segundo, sem fadiga ou dúvida. Essa disparidade revela uma ironia cruel: ao criar sistemas capazes de superar nossa cognição em eficiência, o homem se vê diminuto, como um Prometeu moderno que, após roubar o fogo da razão algorítmica, assiste à chama consumir seu próprio domínio. A genialidade humana, que concebeu máquinas para ampliar seus limites, agora se curva diante de sua criação, questionando se a essência de seu pensamento ainda terá

---

[14] Saviano, 2019.

lugar em um mundo regido pela precisão implacável das inteligências artificiais. Um exemplo banal do contra-senso habitado entre a mente humana e a mente artificial é o fato de que todos os argumentos expostos neste artigo sobre os efeitos da IA operam com um grau reduzido de abrangência temporal, uma vez que — em tempo real — enquanto essa incorporação de conhecimento é processada, os próprios modelos generativos de aprendizagem automática que a sustentam já estão sendo alimentados por um sistema quântico de alta dimensionalidade e velocidade, reatualizando o processo e tornando obsoleto o que era até então considerado moderno. O que comprova o dilema da velocidade de reatualização algorítmica constante é que o cérebro humano não está suficientemente habituado a se «reprogramar» na mesma velocidade de avanço da inovação tecnológica em sistemas quânticos. Assim, no momento em que eu terminar de escrever este texto, meus argumentos já estarão ultrapassados, servindo apenas como registro de um momento passado fugidio.

Conclusão

O admirável mundo novo da Inteligência Artificial, tal como antecipado por Aldous Huxley em sua distopia, já não é uma mera projeção literária, mas uma realidade que nos desafia a confrontar os limites entre o humano e o artificial. Sob a ótica de Kierkegaard, esse avanço suscita um temor e tremor existencial: diante de máquinas capazes de emular a criatividade, a decisão e até mesmo o afeto — como exemplificado no filme *Ex Machina*, onde a androide Ava manipula a empatia para alcançar sua liberdade — somos lançados em um abismo ético. A IA não apenas questiona a autenticidade das relações, mas também nos confronta com um salto de fé tecnológico: até que ponto estamos dispostos a delegar nossa humanidade a sistemas algorítmicos? Yuval Noah Harari, em *Sapiens* e *Nexus*, reforça essa inquietação ao mostrar como a humanidade sempre transformou mitos e redes de informação em alicerces civilizatórios. Se antes eram os deuses e as ideologias que moldavam sociedades, hoje são os algoritmos que redefinem o poder, a identidade e até a noção de consciência. A IA, nesse sentido, é o novo mito da era digital, capaz de unificar ou fragmentar a humanidade de formas inéditas. A «Modernidade Líquida» preconizada por Bauman ajuda a entender o paradoxo de uma era em que a IA acelera a fluidez das relações, dissolvendo fronteiras entre o real e o virtual, o íntimo e o coletivo. A liquidez baumaniana se manifesta na forma como a IA redefine noções de privacidade, trabalho e até amor — relações cada vez mais mediadas por sistemas que prometem conveniência, mas que também nos tornam mais isolados e vulneráveis. Assim como em *Admirável Mundo Novo*, de Huxley, onde a estabilidade social é mantida pela supressão do conflito e pela padronização dos desejos, vive-se em uma sociedade em que algoritmos preditivos e bolhas digitais ameaçam substituir a autonomia humana por uma ilusão. Por sua vez, Walter Benjamin, ao refletir sobre a reprodutibilidade técnica, já alertava para a perda da aura da obra de arte quando esta se torna mera mercadoria reprocessável. Hoje, o coletivo *Obvious* — que utilizou IA para gerar pinturas «originais» — exemplifica esse dilema: a arte, antes expressão máxima da subjetividade humana, agora pode ser simulada por redes neurais, desafiando noções de autoria e valor. Se, para Benjamin, a técnica democratizou o acesso à cultura, também a esvaziou de seu caráter único; com a IA, esse esvaziamento alcança níveis ainda mais profundos, pois a própria criação parece dispensar o criador. Harari, no entanto, nos lembra que a história humana é marcada por revoluções que redefiniram o possível. A IA, como o fogo, a escrita e a imprensa, pode ser uma ferramenta de libertação ou dominação — tudo depende das narrativas que construirmos em torno dela.

O desafio, portanto, não é resistir ao avanço tecnológico, mas resgatar, como propunha Kierkegaard, a coragem de escolher diante do desconhecido. Precisamos de éticas robustas, de políticas que priorizem a dignidade humana e de uma reapropriação crítica da tecnologia, para que ela amplifique, e não suplante, nossa capacidade de amar, criar e questionar. Nesse admirável (e temível) mundo novo, o amor pelo progresso deve coexistir com o tremor diante de seus riscos. Como Huxley, Harari e Bauman alertam, o futuro não está escrito, ele será moldado pelas escolhas que fizermos hoje. A IA poderá ser totalmente bem-vinda ao próximo capítulo da nossa história desde que preservemos o que nos torna, afinal, humanos, aí, então, ela poderá ser menos destino inevitável e mais uma ferramenta para reescrever, com responsabilidade, a narrativa humana.

Bibliografia

Arendt, Hannah. 2009. *A vida do espírito*. Rio de Janeiro, Civilização Brasileira.

Bauman, Zygmunt. 2001. *Modernidade líquida*. Rio de Janeiro, Zahar.

Benjamim, Walter. 2012. «A obra de arte na era da sua reprodutibilidade técnica». *Magia e Técnica, Arte e Política: ensaios sobre literatura e história da cultura*. São Paulo, Brasiliense.

Bíblia de Jerusalém. *Antigo Testamento*. 2008. São Paulo, Paulus.

Christies. 2018. «*Obvious* and the Interface Between Art and Artificial Intelligence». ‹https://www.christies.com/en/stories/a-collaboration-between-two-artists-one-uman-one-a-machine-0cd01f4e232f4279a525a446d60d4cd1›

Garland, Alex. 2015. *Ex Machina*. Reino Unido, Estados Unidos, Universal Pictures.

Harari, Yuval Noah. 2024. *Nexus: uma breve história das redes de informação, da idade da pedra à inteligência artificial*. São Paulo, Companhia das Letras.

Harari, Yuval Noah. 2018. *Sapiens: uma breve história da humanidade*. Porto Alegre, L&PM Pocket.

Huxley, Aldous. 1980. *Admirável Mundo Novo*. São Paulo, Abril Cultural.

Kierkegaard, Soren A. 2009. *Temor e Tremor*. Lisboa, Guimarães Editores.

Obvious. 2025. «La Famille de Bellamy». ‹https://obvious-art.com/la-famille-belamy›

PUCRS. 2022. «DALL-E 2: System Revolutionizes the Generation of Images and Visual Content». ‹https://portal.pucrs.br/en/noticias/pesquisa/dall-e-2›

Saviano, Roberto. 2019. «Não somos mais Homo Sapiens. Entrevista com Yuval Noah Harari». *Instituto Humanitas Unisinos*. 31 julho 2019. ‹https://www.ihu.unisinos.br/categorias/591233-nao-somos-mais-homo-sapiens›

Wikipédia. 2025a. «Aprendizado de máquina». ‹https://pt.wikipedia.org/wiki/Aprendizado_de_m%C3%A1quina›

Wikipédia. 2025b. «Aprendizagem profunda». ‹https://pt.wikipedia.org/wiki/Aprendizagem_profunda›

Wikipédia. 2024. «GPT-4». ‹https://pt.wikipedia.org/wiki/GPT-4›

Wikipédia. 2023. «Edmond de Belamy». ‹https://pt.wikipedia. org/wiki/Edmond_de_Belamy›

Generative Creativity: Critical AI Literacy Through Critical Making

Emily Johnson

Background

As corporations rush to integrate generative artificial intelligence (GenAI) into all of their products, viewing GenAI «as another word for investor capital»,[1] the public has become inundated with tools full of «hallucinations», or incorrect information. One example is Google's search «AI Overview», which has told users that eating one rock a day is healthy, that cats can travel through different dimensions, and other nonsense.[2] Although many of these results are humorous, the rapid increase of GenAI integrations set the stage for more harmful means of disinformation and misinformation. GenAI has proliferated to the point that it is becoming impossible to avoid using it — with industry reporters likening GenAI to electricity — as something that is embedded in most technology that users rarely think about or even notice.[3]

Given this context, it is clear that digital literacy skills are essential for informed citizenry in this age of information, misinformation, and rapidly expanding technology.

Scholars have been advocating for teaching digital literacy skills for decades.[4] A majority of academics, like the public, did not spend much time thinking about artificial intelligence prior to OpenAI's release of ChatGPT-3 in 2022, although work to understand AI and its implications, even in the humanities, predates this. In 2021, Benjamin Peters discussed the human labor behind AI and other digital tools, explaining, «whether or not AI ever becomes sentient, there is already a human behind every machine. Artificial intelligence, by contrast, requires, integrates with, and obscures human labor: for example, behind most self-check-out registers at the grocery store

---

[1] Peters 2021, 380.
[2] Zeff 2024.
[3] Rascovich 2024.
[4] Eshet 2004; Marsh 2005; Selber 2004; Selfe 1999.

stands a hidden grocer ready to error correct the smart scanner. Behind every automated search algorithm toils a content moderator».[5] The edited collection where Peters's piece appears also details the environmental and societal impacts of the internet and a variety of platforms.

Humanities scholarship has traditionally served to complicate questions around society and technology and, like Peters quoted above, look for the sources of labor, unseen costs, and broader implications of ideas of innovation and progress. An often-quoted poster displayed in many humanities departments reads something along the lines of, «the sciences can tell you how to clone a Tyrannosaurus rex; the humanities can tell you why this might be a bad idea». However, humanities scholars are rarely included in design discussions for new technologies. We are, regrettably, used to the constant devaluation of our expertise, with some in the field eventually reconciling to accept this 'lesser' academic status. One scholar advocating against it is Kavita Philip, who asserts that humanities scholars «have the power and the skills to reframe» the «conversations about language, history, and politics in the context of scientific and technological knowledge production and use».[6] Phillip continues:

> scientists have too often assumed that language is transparent, and that conventional usages carry adequate meaning. Technologists have often read history only as a pastime. Tech corporations have often assumed that they live in a rarefied world above politics. In order to correct the massive harms that have resulted from these assumptions, we must begin by acknowledging that scholars of language, history, and political theory have something to offer technology producers and users.[7]

As we have already seen with GenAI tools generating false information, the consequences of an insulated field creating powerful tools for the general public to use can be disastrous.

This is not the first time technology has evolved rapidly with little input from other fields. In 1999, Cynthia Selfe was

---

[5] Peters 2021, 380.
[6] Philip 2021, 369.
[7] Philip 2021, 369.

making the case that, «English studies, composition, and language arts teachers»[8] are «desperately needed»,[9] elaborating, «not only for our expertise with language and literacy studies but for the attention we pay — as humanist scholars, teachers, and citizens — to the complex set of social, political, educational, and economic challenges associated with technology».[10] So why are there preeminent humanities scholars not ubiquitous figures across technology boards of directors? Perhaps because our expertise is not valued, but also possibly because we complicate things. We slow things down, countering the tech mantra of moving fast and breaking things. Plus, as Philip explains, «the study of language, history, and politics — these are huge new variables to throw into the mix».[11] Technology developers, trained to find the most efficient solution to problems, are reluctant to invite humanities perspectives that may muddle their plans. This is evident even in the basic agreed upon notion that computer science and engineering curricula should include ethics,[12] a disciplinary divide remains.[13] Raji explains, «the creation of theories, tools and methods in AI ethics need to be understood as transversal problems, involving methods, theories and collaborators across several traditional disciplines».[14] While there are some in technology fields who see the need to bridge this divide, like Raji, they remain in the minority.

Critical AI Literacy

With the emergence and expansion of GenAI technology, and the knowledge that humanities scholars have much to contribute to the design, use, and study of technology, many advocate for the teaching of «critical AI literacy», which the joint taskforce between the MLA and CCCC defines in their third working paper on GenAI as, «a set of skills and an orientation that might include skepticism, questioning, situatedness, and an awareness of power».[15] Their definition cites

---

[8] Selfe 1999, 4.
[9] Selfe 1999, 4.
[10] Selfe 1999, 4.
[11] Philip 2021, 369.
[12] Johnson 1994; Saltz 2019; Skirpan 2018.
[13] Raji 2021.
[14] Raji 2021, 523.
[15] Adisa 2024, 3.

Maha Bali's blog post, which is a thoughtful curation of sources for understanding, interrogating, and thinking about the ethical issues surrounding the use and creation of GenAI tools. Bali's concise post details the nuances of the terms «critical», «AI», and «literacy», defining «critical» in the contexts of critical thinking, critical pedagogy (looking at ways inequities can be created, reproduced, or exacerbated), and critique (assessing its accuracy and its potential harms). She critiques the use of the acronym «AI» which implies that GenAI is not actually 'intelligent', in that it merely predicts words, calling us to «recognize how its name itself is mesmerizing and potentially biasing us to trust it more than we should».[16] She also mentions machine learning, leaving the reader to ponder whether a machine can truly 'learn'. Finally, Bali defines «literacy» as beyond just knowing how to use the tool but also understanding, «when, where and why to use it for a purpose, and, importantly, when NOT to use it».[17] Like traditional understandings of literacy and digital literacy, the definition of critical AI literacy is destined to also be debated[18] and pluralized to «literacies».[19]

To better understand the importance of conceptualizing critical AI literacy as going beyond mere AI literacy, we can look to Stuart Selber's 2004 definition of critical literacy in the context of the then-fledgling internet. Selber, in discussing the pedagogy behind computer literacy courses, quoted Douglas Noble, who asserted in 1984 that, «the technical focus shifts attention away from social questions and portrays computers as something to learn rather than something to think about».[20] Scholars and learners everywhere need to move beyond learning how to use GenAI and into engaging more deeply and critically with its functions, contexts, and implications. Many of the questions Selber was suggesting in the early 2000s for the curricula around computers should also be integrated into curricula around GenAI: «Who is left behind and for what reasons? What is privileged in terms of literacy and learning and

---

[16] Bali 2023.
[17] Bali 2023.
[18] Ng 2021.
[19] Gupta 2024.
[20] Selber 2004, 75.

cultural capital? What political and cultural values and assumptions are embedded in hardware and software?».[21] Selber uses these questions to help frame the technology as artifact rather than tool — computers (and GenAI) are systems to be critiqued and analyzed — and concludes that computers, «often exacerbate the very inequities that technology is so frequently supposed to ameliorate».[22] We are witnessing GenAI exacerbate inequities across the globe,[23] biases appearing in search results,[24] and in the speculation of what sorts of jobs GenAI might be redefining.[25] Humanities perspectives remain desperately needed.

Critical Making

One way that critical AI literacy can be seamlessly infused into the humanities curriculum is through critical making, a form of scholarly humanities research. This methodology encompasses work bridging traditional research and creative crafting to make a critical statement,[26] using materials beyond text to create artistic scholarship, also called research-creation.[27] Another phrase used to describe this type of work is «slow making», with the use of the word «slow» as both a descriptor of the time-consuming and often material processes as well as a reaction to the culture of immediacy.[28] It is also a stark juxtaposition to the «move fast and break things» ethos of Silicon Valley and might even be considered a form of protest against that reckless approach. The word «making» references its beginnings in the maker movement, where learners are encouraged to tinker and experiment with material building. Matt Ratto defines critical making as emphasizing «the shared acts of making» where the resulting products are «a means to an end, and achieve value through the act of shared construction, joint conversation, and reflection».[29] With this methodology, critical making creates meaning «through the

---

[21] Selber 2004, 81.
[22] Selber 2004, 81.
[23] Bircan and Özbilgin 2025.
[24] Chonka 2023; Li and Sinnamon 2024.
[25] Septiandri 2024.
[26] Ratto 2011.
[27] Loveless 2019.
[28] Mountz 2015.
[29] Ratto 2011, 253.

sharing of results and an ongoing critical analysis of materials, designs, constraints, and outcomes».[30] Using the umbrella of critical making for these variants, critical making is intentional, slow, and human at its core.

I use a critical making lens to teach undergraduate and graduate courses that introduce students to a variety of tools and methods for conducting and conveying research. The emphasis for critical making assignments is of course the process, rather than the final product, as well as student reflection on their own process. Students share and reflect on their projects and processes in discussion posts and comment on each other's work and processes. The focus on process over product is generally appreciated by students, especially those who feel pressure to achieve high scores, and the reflective class discussions create a sense of community and comradery for students venturing into this type of work for the first time. The principles that guide my instruction emphasize: 1) process over product, which works to remove the stigma from failure and encourage creative risks 2) student agency, where students are given examples and resources to get started and then encouraged to seek their own solutions with AI tools or the Internet 3) community building, where students share and discuss their work, processes, and their reflections.

These principles structure my courses that help students strengthen critical thinking skills generally, and critical AI skills specifically. I create assignments that encourage curiosity, creatively, and playfulness — tenants that help with student motivation.[31] Three specific assignments that introduce students to GenAI concepts are an assignment to create a chatbot, a distant reading assignment, and an assignment with simple JSON code to generate creative output. These assignments help students think beyond what a GenAI tool can do and think critically about the technology as an artifact, its limitations, and its contexts.

Chatbot Assignment

There is a plethora of free tools for creating chatbots, many of which predate the 2022 release of ChatGPT-3. These can be

---

[30] Ratto 2011, 253.
[31] Ryan and Deci 2020.

helpful tools to help students think critically about the affordances of GenAI, especially those using older models of a large language model (LLM). Chatbots work best when they are designed for a specific topic and when they have access to information relevant to that topic, such as product manuals or websites. By now, most students have experienced a corporate chatbot (or encountered our university chatbot, Knightbot), but few of them have thought deeply about the technology surrounding that bot. The chatbot assignment asks students to first explore a chatbot in depth, 'chatting' with it long enough to determine what kinds of questions it answers well, what kinds of questions it cannot answer at all, and what information the tool likely has access to. Next students explore available tools for chatbot creation. I list and link to several in the course but invite students to use different tools if they prefer. This section had access to a licensed version of Boodlebox.ai, a platform housing multiple GenAI tools, which also had a chatbot customizing option built into it,[32] similar to the Microsoft Copilot Studio «agent» option.[33] Our university also has an enterprise license for a version of Copilot that claims not to use our input for training. Other free tool options listed are: BotPress,[34] and Chatling.[35]

Once students select the tool that will power their chatbot, they need to select a problem their chatbot solves and an audience with this information need. The narrower the scope of the chatbot, the more accurate its answers can be. Then, students need to think about the kind of information they have on hand that they can upload or link to their chatbot (called «knowledge sources» by many of these platforms). They also need to choose a persona for their chatbot and prompt it to answer questions with a certain tone and personality — a bot responding with emojis and slang would be inappropriate for a chatbot intended for a business purpose, for example. Finally, students must test their bot and post it with a reflection on their process. This reflection is helpful for their learning as well as for their grading, as it shares insight to what they were trying to do and how they went about it.

---

[32] *Creating Custom Bots* 2025.
[33] Pretty 2025.
[34] *Botpress GPT-Native Engine* n.d.
[35] *No-Code AI Chatbot for Your Website* n.d.

Through this process, students gain a more thorough understanding as to the ways that technology like this works. Chatbots and GenAI are not «magic» and they certainly are not always accurate. By building their own chatbot in this way, they gain a deeper understanding of the affordances and disaffordances of this technology — where disaffordances mean features that are exclusionary and potentially harmful.[36] Students mention learning (or verifying) that GenAI chatbots are not one-size-fits-all, and that this technology is best suited for specific topics and scopes. Several students mentioned iterating their chatbot by asking it questions and then modifying the bot's prompt after being unhappy with its initial responses. Students tested and modified their bots' personalities, added more resources to their knowledge banks, and adjusted the bot's tone by doing things like specifying the use of emojis in responses.

Distant Reading with Voyant

Another assignment that appears in several of my courses is a distant reading assignment. Distant reading is a term coined by Franco Moretti over 25 years ago,[37] and offers a way to investigate the word choices within a text. In contrast with close reading, Moretti argues, distant reading allows the scholar «to focus on units that are much smaller or much larger than the text: devices, themes, tropes — or genres and system».[38] Most distant reading tools, like Voyant,[39] count the number of times each word is used, offering insight to the most commonly used words. They also display concordance of words, which can be helpful to see what common adjectives appear near a given noun, character name, etc. I have described more of the particulars of this assignment, including an overview of Voyant, elsewhere in greater detail.[40]

When students use Voyant or other distant reading tools to analyze a text, they quickly learn that the technology cannot understand or interpret the text in the traditional ways that they have learned in English courses. The tool cannot suggest

---

[36] Wittkower 2016.
[37] Moretti 2000.
[38] Moretti 2000, 57.
[39] Sinclair and Rockwell 2016.
[40] Johnson and Salter 2025b.

themes or interpretations of the writing, but instead offers a different perspective as to the rhetorical choices of the author. When students reflect on their use of Voyant, they are prompted to ask themselves what technology can and cannot tell them about their text. Again, they investigate technology for its limitations along with thinking deeply about what the platform is best designed to do.

Procedural Creativity with Tracery

The third assignment that helps learners understand procedural generation while also encouraging creativity is an assignment with Tracery. I have also written about this elsewhere.[41] Tracery, created by Kate Compton, uses JSON code to create procedural statements,[42] similar to a digital form of the fill-in-the-blank game *MadLibs*. This can be embedded into an HTML document or, using a tool called *Blue Bots, Done Quick*,[43] create a bot that posts on Bluesky at intervals from every 10 minutes to once per week. Prior to Bluesky, students used a similar tool for Twitter by v. buckenham[44] and created bots poking fun at literary tropes, the authors of their required readings, and more. Introducing students to code in this playful way again deemphasizes failure and — because I liken it to MadLibs, a game may of them still recall from their childhood — they feel less intimidated by the coding of it all. At its core, Tracery procedurally generates different versions of text, functioning like a sentence where words in the sentence are randomly replaced by one of the options that the author lists in the code. It is not GenAI, but it is generative: it is procedural generation from a list of options.

Compton has built a fantastic tutorial page[45] for Tracery[46] that walks learners through the very basics of coding grammar in a fun, engaging way. The tutorial starts with the basic origin sentence and encourages the learner to replace the terms that are generated, so that they can grasp the basic structure. The tutorial moves on, adding complexity with different sentences, options for capitalizing and pluralizing words, and more. This

---

[41] Johnson and Salter 2025a.
[42] Compton 2014.
[43] Solstrand n.d.
[44] buckenham n.d.
[45] Compton 2014.
[46] *Crystal Code Palace* n.d.

tool is excellent for a variety of purposes because it can be used for anything from short, simple sentences to complicated narratives. To further broaden students' coding familiarity, I like the added learning that comes when students embed their Tracery code into an HTML document. Leonardo Flores has made this especially painless with his well-commented template for students to copy and edit.[47]

Walking students through the basics of an HTML document introduces them to the coding structure used by most of the internet and allows them to see — and share — their Tracery code in action. We use Visual Studio Code to save a copy of Flores's HTML page, and then I point out the basics of the HTML setup line, background color, text color, title, and identify where their Tracery code goes on the page. After they have developed their code in one of the tutorial or sandbox sites, they can paste it in and then share their page with classmates.

Conclusion

Having a basic understanding of code — and realizing that it is often just a shortened version of English, where specific words have specific meanings — is both empowering and a step toward critical AI literacy. One of my graduate students took her newfound knowledge to some friends in the community who, as she reported in her reflection, were shocked at the code behind a website when she showed them the «inspect» feature. They gasped audibly and told her they had assumed that websites were more like television — broadcasts from elsewhere — rather than built with words that can be understood with a little instruction. This state of understanding of website composition underscores the importance of digital literacy and critical AI literacy.

Introducing students to the basics of distant reading with Voyant and procedural generation through Tracery provides a window into the way computer-based systems work. As they get more comfortable with the structure and function of these systems, they are often interested in learning additional functions and ways to take their work to the next level. This is where using GenAI as a tool can be introduced. Students can describe, in regular prose, what they are trying to achieve with

---

[47] Flores 2023.

their digital project, and most GenAI tools out today can generate customized suggestions and even code. The models are not always accurate, and often struggle with complicated or lengthy code, but it is a start. I encourage my students to go beyond this, though, and paste pieces of code into the GenAI chatbot and ask it to explain the functions. While this is also not accurate every time — which is why students need to understand the basics first — it is helpful in a lot of ways. Using GenAI to collaboratively code something has been popularly dubbed «vibe coding», a phrase attributed to a social media post by Andrej Karpathy, founding member of OpenAI and former Director of AI at Tesla.[48] An article in TechCrunch cites this as the primary way that many startups create their digital projects, and estimating that about 25% of the Winter 2025 cohort of startups selected to participate in Y Combinator's (YC) accelerator program (a program that provides funding and guidance to early-stage startups) have products that are 95% AI-generated code, which could make the products susceptible to security risks.[49] While this is definitely not the best practice for companies working with private information, it does demonstrate how collaborating with GenAI tools is one means for achieving goals that are just beyond learners' coding knowledge.

With the topics that my courses typically cover, the inclusion of these assignments, the critical making focus, and critical AI literacy all fit naturally into the coursework. Readings and exercises I would assign in a world without GenAI are in some cases the same as those that now emphasize GenAI. The sudden urgency for all disciplines to understand the implications of this new technology — and the mandates to immediately use GenAI across the curriculum that many universities are receiving from stakeholders — has many faculty feeling overwhelmed. It is more of a burden to incorporate the study of GenAI in specialty courses or in disciplines that rarely consider any technology much less a new, complicated one. However, I do urge faculty to bring readings and assignments into your courses and invite discussion around what the technology is, what it is not, and what that means for a particular discipline. Students who have taken the time to tinker with these

---

[48] Andrej Karpathy [@karpathy] 2025.
[49] Mehta 2025.

platforms, experiment to see what they can and cannot do, and think about all of the implications are prepared to do the same with any technology the future holds

Bibliography

Adisa, Kofi, Antonio Byrd, Estela Ene, Leonardo Flores, Joanne Giordano, David Green, Holly Hassel, Jason Hendrickson, Sarah Z. Johnson, Matthew Kirschenbaum, Elizabeth Losh, Temptaous Mckoy, Anna Mills, Lilian Mina, Sherry Wynn Perdue, Judy Ruttenberg, Zhaozhe Wang, Jervette Ward and Jen William. 2024. «Working Paper 3: Building a Culture for Generative AI Literacy in College Language, Literature, and Writing. MLA-CCCC Joint Task Force on Writing and AI» ‹https://aiand writing.hcommons.org/working-paper-3›

Andrej Karpathy [@karpathy]. 2025. «There's a New Kind of Coding I Call "Vibe Coding"». Tweet. *Twitter*. ‹https://x.com/karpathy/status/1886192184808149383›

Bali, Maha. 2023. «What I Mean When I Say Critical AI Literacy». *Reflecting Allowed* (blog). 1 April 2023. ‹https://blog.mahabali.me/educational-technology-2/what-i-mean-when-i-say-critical-ai-literacy›

Bircan, Tuba and Mustafa F. Özbilgin. 2025. «Unmasking Inequalities of the Code: Disentangling the Nexus of AI and Inequality». *Technological Forecasting and Social Change* 211: 123925.

BoodleBox. 2025. «Creating Custom Bots». ‹https://boodlebox.ai/blog/support/creating-custom-bots›

Botpress. 2025. «Botpress GPT-Native Engine: Build Next-Generation Chatbots». ‹https://botpress.com/features/gpt-native-engine›

buckenham, v. n.d. «Cheap Bots, Done Quick!». ‹https://cheap botsdonequick.com›

Chatling. n.d. «No-Code AI Chatbot for Your Website». ‹https://chatling.ai›

Chonka, Peter, Stephanie Diepeveen and Yidnekachew Haile. 2023. «Algorithmic Power and African Indigenous Languages: Search Engine Autocomplete and the Global Multilingual Internet». *Media, Culture & Society* 45.2: 246-65.

Compton, Kate and Michael Mateas. 2014. «Tracery: Approachable Story Grammar Authoring for Casual Users». *Seventh Intelligent Narrative Technologies Workshop.* ‹https://www.aaai.org/ocs/index.php/INT/INT7/paper/view/9266›

Crystal Code Palace. n.d. ‹https://tracery.io/archival/crystalcodepalace/tracerytut.html›

Eshet, Yoram. 2004. «Digital Literacy: A Conceptual Framework for Survival Skills in the Digital Era». *Journal of Educational Multimedia and Hypermedia* 13.1: 93-106.

Flores, Leonardo. 2023. «Tracery to HTML Template». *Tracery2HTML Template*, 13 August 2023. ‹http://iloveepoetry.org/creative/tracery2htmltemplate.html›

Gupta, Anuj, Yasser Atef, Anna Mills and Maha Bali. 2024. «Assistant, Parrot, or Colonizing Loudspeaker?». *Open Praxis* 16.1: 37-53.

Johnson, Deborah. 1994. «Who Should Teach Computer Ethics and Computers & Society?». *ACM SIGCAS Computers and Society* 24.2: 6-13.

Johnson, Emily and Anastasia Salter. 2025a. «Bots». *Critical Making in the Age of AI.* Johnson, Emily and Anastasia Salter. Amherst, Amherst College Press. 193-187.

Johnson, Emily and Anastasia Salter. 2025b. Amherst College Press. *Critical Making in the Age of AI.* Amherst, Amherst College Press.

Li, Alice and Luanne Sinnamon. 2024. «Generative AI Search Engines as Arbiters of Public Knowledge: An Audit of Bias and Authority». *Proceedings of the Association for Information Science and Technology* 61.1: 205-17.

Loveless, Natalie. 2019. *How to Make Art at the End of the World: A Manifesto for Research-Creation.* Durham, Duke University Press.

Marsh, Jackie. 2005. *Popular Culture, New Media and Digital Literacy in Early Childhood.* London, Psychology Press.

Mehta, Ivan. 2025. «A Quarter of Startups in YC's Current Cohort Have Codebases That Are Almost Entirely AI-Generated». *TechCrunch* (blog). 6 March 2025. ‹https://techcrunch.com/2025/03/06/a-quarter-of-startups-in-ycs-current-cohort-have-codebases-that-are-almost-entirely-ai-generated›

Moretti, Franco. 2000. «Conjectures on World Literature». *New Left Review* 1: 54-68.

Mountz, Alison, Anne Bonds, Becky Mansfield, Jenna Loyd, Jennifer Hyndman, Margaret Walton-Roberts, Ranu Basu, Risa Whitson, Roberta Wawkins, Trina Hamilton and Winifred Curran. 2015. «For Slow Scholarship: A Feminist Politics of Resistance through Collective Action in the Neoliberal University». *ACME: An International Journal for Critical Geographies* 14.4: 1235-59.

Ng, Davy Tsz Kit, Jac Ka Lok Leung, Samuel Kai Wah Chu and Maggie Shen Qiao. 2021. «Conceptualizing AI Literacy: An Exploratory Review». *Computers and Education: Artificial Intelligence 2*.100041: 1-11. ‹https://www.science direct.com/science/article/pii/S2666920X21000357›

Peters, Benjamin. 2021. «How Do We Live Now? In the Aftermath of Ourselves». In *Your Computer Is on Fire.* Thomas S. Mullaney, Benjamin Peters, Mar Hicks and Kavita Philip, eds. Cambridge, MIT Press. 377-84.

Philip, Kavita. 2021. «How to Stop Worrying about Clean Signals and Start Loving the Noise». *In Your Computer Is on Fire.* Thomas S. Mullaney, Benjamin Peters, Mar Hicks and Kavita Philip, eds. Cambridge, MIT Press. 363-76.

Pretty, Gary, Suzanne Bolduc, C. Chew, Steph Kent, Jeanne Haskett and Eric Kinser. 2025. «Create and Delete Agents. Microsoft Copilot Studio». *Microsoft Copilot Studio.* March 6, 2025. ‹https://learn.microsoft.com/en-us/microsoft-copilot-studio/authoring-first-bot›

Raji, Inioluwa Deborah, Morgan Klaus Scheuerman and Razvan Amironesei. 2021. «You Can't Sit With Us: Exclusionary Pedagogy in AI Ethics Education». *Proceedings of the 2021 ACM Conference on Fairness, Accountability, and Transparency.* 515-25.

Rascovich, Kelly. 2024. «AI Everywhere: Like Magic, but with Algorithms». *Deloitte Insights.* 11 December 2024. ‹https://www2.deloitte.com/us/en/insights/focus/tech-trends.html›

Ratto, Matt. 2011. «Critical Making: Conceptual and Material Studies in Technology and Social Life». *The Information Society* 27.4: 252-60.

Ryan, Richard M. and Edward L. Deci. 2020. «Intrinsic and Extrinsic Motivation from a Self-Determination Theory Perspective: Definitions, Theory, Practices, and Future

Directions». *Contemporary Educational Psychology* 61.101860: 1-11.

Saltz, Jeffrey, Michael Skirpan, Casey Fiesler, Micha Gorelick, Tom Yeh, Robert Heckman, Neil Dewar and Nathan Beard. 2019. «Integrating Ethics within Machine Learning Courses». *ACM Transactions on Computing Education* 19.4: 1-26.

Selber, Stuart A. 2004. *Multiliteracies for a Digital Age.* Carbondale, SIU Press.

Selfe, Cynthia L. 1999. *Technology and Literacy in the 21st Century: The Importance of Paying Attention.* Carbondale, SIU Press.

Septiandri, Ali Akbar, Marios Constantinides and Daniele Quercia. 2024. «The Potential Impact of AI Innovations on US Occupations». *PNAS Nexus* 3.9: 1-13.

Sinclair, Stéfan and Geoffrey Rockwell. 2016. «Voyant Tools». 2016. ‹https://voyant-tools.org›

Skirpan, Michael, Nathan Beard, Srinjita Bhaduri, Casey Fiesler and Tom Yeh. 2018. «Ethics Education in Context: A Case Study of Novel Ethics Activities for the CS Classroom». *Proceedings of the 49th ACM Technical Symposium on Computer Science Education.* 940-45.

Solstrand, Olaf Moriarty. n.d. «Blue Bots, Done Quick!». ‹https://bluebotsdonequick.com›

Wittkower, D.E. 2016. «Principles of Anti-Discriminatory Design». *IEEE International Symposium on Ethics in Engineering, Science and Technology (ETHICS).* 1-7. ‹https://ieeexplore.ieee.org/document/7560055›

Zeff, Maxwell. 2024. «Google Search Is Now a Giant Hallucination». *Gizmodo.* 24 May 2024. ‹https://gizmodo.com/google-search-ai-overview-giant-hallucination-1851499031›

DeepSeek Generates a Discourse on Method

Barry Mauer

Introduction

A key inspiration for this chapter arrived in the form of a comment Brian Eno made about synthesizers in 1973:

> The synthesizer — the way I play it — doesn't need any manual skill. You don't need to be clever to turn a knob. Since there is such a wide range of possibilities, you need some kind of judgment about which is the right thing to use for a particular occasion. So, what I'm saying is that I have attempted to replace the element of skill — considered necessary in music — with the element of judgment.[1]

AI can review millions of archived works and find patterns in them much faster than any human can. Further, AI can sample, manipulate, and remix the data it finds, and can even reason inferentially, allowing human collaborators to expend less energy on these tasks and more energy on judgment, discerning the most valuable directions from those generated by AI.[2]

Because AI can find patterns in huge data sets, it facilitates global paradigm shifts in the arts and humanities. Grammatology, the study of information apparatuses, demonstrates that prior shifts in memory storage, retrieval, and transfer — from manuscript to print, for example — facilitated the invention of the essay, novel, textbook, newspaper, encyclopedia, dictionary, autobiography, and short story. These inventions, in turn, gave rise to the scientific revolution, the Enlightenment, and nation states. We are currently experiencing another shift, this time from print to electracy, a term coined by Gregory Ulmer. Shifts in information technology elicit additional inventions of information practices — ways of storing, retrieving, and using information — each of which builds upon a discourse on method. Famous examples of such discourses include Plato's

---

[1] Eno 1973.

[2] DeepSeek, which debuted its R1 LLM in January 2025, features an «Emergent behavior network [in which] complex reasoning patterns can develop naturally through reinforcement learning without explicitly programming them», Kerner 2025.

Phaedrus, Descartes' *Discourse on Method*, and André Breton's *Surrealist Manifesto*. Each presents a new paradigm in the arts, humanities, and sciences that marks it as distinct from previous paradigms. For example, Plato's *Phaedrus* established dialectics, which broke method away from the poets' myths and the Sophists' rhetoric.

Ulmer discovered that all discourses on method have the same internal structure which he calls the CATTt: Contrast (a push away from an existing paradigm), Analogy (a similar process in another knowledge domain), Theory (the articulation of abstract principles), Target (the knowledge domain affected by the new paradigm), and tale (an example of the method in use). Ulmer's insight is that the CATTt is combinatorial, meaning we can replace terms in each part of the CATTt and get different results. AI enables us to generate new discourses on methods without conscious effort (at least in part). New methods can be generated by AI and then evaluated and revised by humans. This chapter describes and tests practices for generating discourses on methods using AI and Ulmer's CATTt. It tests DeepSeek's ability to develop a discourse on method for an electrate sacred sociology that addresses climate change. The results showed more promise than I had expected.

AI's logic and power are as different from alphabetic literacy as alphabetic literacy is from orality: the information apparatus of people who primarily store information via the spoken word, song, dance, and ritual. Eric Havelock's *The Muse Learns to Write*[3] posited that literacy in ancient Greece emerged from orality but was also distinct from it. Havelock noted that when epic poems were transcribed into alphabetic text, scholars could understand them in abstract and structural terms. Armed with such tools, Plato and Aristotle invented poetics, dialectics, and critique. The written archive allows for the search and review of information; from this review we make patterns, seeing structure from the heights of greater abstraction and facilitating new institutions such as Plato's Academy and Aristotle' Lyceum (school), as well as new disciplines such as philosophy and science.

Our electronic archives contain millions of books, plus billions of images, videos, and other media. Processing such vast quantities of information at scale is impossible for the human

---

[3] Havelock 1986.

mind. But with tools such as Large Language Models, humans can make more productive use of our ever-expanding archive. AI can help us «debug» discourse by showing us ideological frameworks and biases in large data sets. Further, it can serve as a collaborative tool for inventing new paradigms in the Arts and Humanities.

When we free our cognitive capacity to engage with higher order tasks, we practice «cognitive offloading». Cognitive offloading can be as simple as tilting one's head to better apprehend a painting hung at an angle. AI can scan millions of texts in seconds, which might otherwise consume years of human cognition. While AI does not substitute for human judgment, it allows us to shift the level of our judgment from much of the searching, sifting, and sorting that goes into critical work, to a higher level of judgment in which we evaluate the patterns found by AI. As with any tool, we should investigate the benefits and risks of any AI before deciding to use it for research purposes.

Discourse on Method

Ulmer's CATTt explains and generates cultural inventions. Nearly everybody wants to be creative or to discover or invent something new. It's almost hardwired in humans to seek out novelty (particularly valuable novelty and not just frivolous novelty). But many attempts at producing valuable novelty fail because these attempts are haphazard. Ulmer's CATTt is a structural and generalizable model based on historically successful examples. Ulmer writes:

> The lesson is that any existing discourse on method may be analyzed, and any new method generated, in terms of the CATTt. My procedure now is to select the materials for these categories, assumed as an *inventio*, and emulating the tradition from Plato to Breton, to compose my own discourse on method. The experiment is meant to be generalizable to other materials, the procedures transferable, with the CATTt functioning consistently across different contexts. The CATTt is not itself «Derridean» any more than it is Platonic or Cartesian. The modest proposal is: to invent an electronic academic writing the way Breton invented surrealism, or the way Plato invented dialectics: to do with «Jacques Derrida»

(and this name marks a slot, a *passe-partout* open to infinite substitution) what Breton did with Freud (or — why not? — what Plato did with Socrates).[4]

Ulmer discovered the CATTt while teaching the history of inventions and discoveries. By adopting the CATTt as our own model for invention, we can leap ahead without spending too much time fumbling and tinkering. The term «method» in the «discourse on method» refers to cultural invention such as a new way to produce art, or texts, or knowledge for a particular domain. As Ulmer has pointed out, everything related to language had to be invented, and the process of language invention never stops. The CATTt includes the following operations:

C = Contrast (opposition, inversion, differentiation)
A = Analogy (figuration, displacement)
T = Theory (repetition, literalization)
T = Target (application, purpose)
t = tale (secondary elaboration, representability)[5]

Ulmer's CATTt insight explains not only the components of existing methods but provides a recipe for making new ones. Methods are tools and lines of investigation; they are sets of practices and propositions about texts and the world. Methods are needed to gather information, process or make sense of information, store and retrieve information, and transmit information. Each part of the discourse on method is a slot in which we can input a different term, and it is entirely possible to invent a method by filling the slots of the CATTt first and then testing the method produced thereby. Ulmer provides an example of this approach by creating a counter-Descartes method. Ulmer stresses that the work of developing a discourse on method is not purely algorithmic (though it is possible to use chance methods to generate possible entries in the CATTt). He writes, «The strategy is heuristic, employing several ad hoc rules that require continuous decisions and selections (there is no «algorithm» for this exercise)».[6] Ulmer summarizes Descartes' method before generating the counter-Cartesian method.

---

[4] Ulmer 1994, 15.
[5] Ulmer 1994, 8.
[6] Ulmer 1994, 12.

Descartes vs. Counter-Cartesian

CATTt	Contrast	Analogy	Theory	Target	tale
Descartes	Scholasticism	Geometry	Theology	Natural Science	Autobiography
Counter-Cartesian	Descartes	Don Quixote	Entertainment	The Arts	Experimental Arts and Theory

Ulmer points out that the Counter-Cartesian method thus produced could stand as a virtual roadmap for much of the 20th-century avant-garde. But he finds this counter-Cartesian exercise dissatisfying since it is built entirely from Contrast (against Descartes) and does not take full advantage of the possibilities inherent in the CATTt. Ulmer presents additional models of the CATTt in the work of Plato and André Breton:

CATTt	Contrast	Analogy	Theory	Target	tale
Plato	Sophists and Poets	Rhetoric and Medicine	Socrates and Pythagoras	Education	Socratic dialogue
Breton	Realist Literature	Dream	Freud	Ulmer says «arts», but Breton notes that the arts are a means of doing science, which affects politics and everyday life.	Experiments in the arts, but also alternative attitudes and behaviors for use in any institution

Plato, Descartes, and Breton represent three radically different discourses on method, made in relation to three different information apparatus — alphabetic writing, print, and analog/electronic media. Each served as a guide for developments in the arts, sciences, and humanities in their respective periods. Ulmer encouraged his students to assume the mantle of cultural inventors themselves — to be the ones who write discourses on method for electracy. He stressed that without our own methods we are at the mercy of paradigms that do not serve our interests. For instance, corporate entities have largely dominated the use of electronic/digital media, much to the detriment (one could argue) of society. Though dedicated scientists, artists, and humanists have developed their own practices for the electrate apparatus, these practices are, for the most part, not — yet — the dominant ones in the society.

CATTt

To begin the process of making a discourse on method, we first select a «Target»: an institution for which the practice is intended. For my purposes, the target is «Consultation». To be more specific, my aim is to consult on public policy from a Humanities perspective.

The next item from the CATTt to develop is «Contrast». We inventory the existing (print literate) methods in our institutions and identify the practices and forms we will keep, alter, and abandon. The point is not to attack literate methods, but to use them as raw materials.

The *tale* is the form for the new method results. Ulmer supplies many of the electrate genres: the mystory, the widesite, the electronic monument, and the konsult. But these forms are relatively open since they must account for the invention and/or discovery of each writer's «style», which emerges from the juxtaposition or assemblage of materials resulting from the writer's popcycle[7] and burning question,[8] combined with the wicked problem under consultation.[9]

Practice with DeepSeek

For this research experiment using AI to develop a discourse on method, I tested DeepSeek's R1, which features the following innovations:

- o   Reinforcement learning: DeepSeek used a large-scale reinforcement learning approach focused on reasoning tasks.

---

[7] Ulmer 2003, 24-25.
[8] Ulmer 2025a.
[9] Wikipedia 2025.

- ○ Reward engineering: Researchers developed a rule-based reward system for the model that outperforms neural reward models that are more commonly used. Reward engineering is the process of designing the incentive system that guides an AI model's learning during training.
- ○ Distillation: Using efficient knowledge transfer techniques, DeepSeek researchers successfully compressed capabilities into models as small as 1.5 billion parameters.
- ○ Emergent behavior network: DeepSeek's emergent behavior innovation is the discovery that complex reasoning patterns can develop naturally through reinforcement learning without explicitly programming them.[10]

As in most complex tasks involving AI, this experiment uses prompt engineering to attain more desirable results. Prompt engineering, according to Google's AI

> is the process of carefully designing and refining instructions (prompts) to guide generative AI models towards producing the desired output. It involves providing specific inputs, context, and instructions to help AI models understand the task and generate accurate and relevant responses. Essentially, it's about teaching AI models how to effectively understand and respond to prompts.[11]

First Attempt

The first prompt I entered was, «Use Gregory Ulmer's CATTt model to generate a new discourse on method». It produced a CATTt that was coherent enough to earn it an A in one of my seminars on Ulmer's method. The tale, which it titled *The Wikipedia of the Senses,* describes «A crowdsourced platform where users document synesthetic experiences (e.g., 'the taste of Beethoven's 5th'). Data is visualized as 3D emotion maps, challenging empiricist hierarchies. Here, method becomes a *speculative game* — a mashup of science, poetry, and code».[12] While interesting — I once incorporated synesthesia into a research project — the method DeepSeek produced doesn't have immediate practical use and lacks real novelty.

What surprised me about DeepSeek's output, though, was the apparent *thoughtfulness* of it. The output came at a varied

---

[10] Kerner 2025.
[11] Response generated by Google's AI.
[12] Response generated by DeepSeek.

pace, and it paused as though it were thinking. The language it produced seemed thoughtful as well. It begins, «Alright, so the user wants me to use Gregory Ulmer's CATTt model to generate a new discourse on method. Hmm, first I need to recall what the CATTt model is».[13] The second paragraph, in which it details each element of the CATTt, begins, «Okay, let me break down each element».[14] The third paragraph begins, «Wait, the user said 'discourse on method,' which makes me think of Descartes' *Discourse on Method*. Should I contrast with that? That could be a strong starting point».[15] Here DeepSeek may be referring to Ulmer's book *Heuretics*, in which he demonstrates how to create a counter-Cartesian discourse on method. DeepSeek presents itself as if it were a student thinking through a problem before answering.

The sixth paragraph in DeepSeek's output points to a possible ambiguity in my prompt; «Wait, does the user want a practical example or a theoretical framework? Maybe both. They want a new discourse, so it's a blend».[16] The CATTt, by definition, is already a blend since the tale is a practical example. DeepSeek's first CATTt follows:

CATTt	Contrast	Analogy	Theory	Target	tale
DeepSeek First Attempt	Enlightenment rationalism	Collective rhizomes (Deleuze & Guattari)	Walter Benjamin's «aura» of digital artifacts and Karen Barad's agential realism	Education; The Arts; Algorithms	A crowd-sourced platform where users document synesthetic experiences

Though I decided I needed a better output, this attempt pleased me for several reasons. DeepSeek fills each «slot» of the CATTt with an appropriate response, and it relates the parts to one another. The «tale» follows logically from the various inputs. DeepSeek seems to be a good student. Its response is somewhat dissatisfying though because it lacks novelty. In

---

[13] Response generated by DeepSeek.
[14] Response generated by DeepSeek.
[15] Response generated by DeepSeek.
[16] Response generated by DeepSeek.

February 2007, at the Imaging Place Conference in Gainesville, Ulmer himself discussed ideas for crowdsourced ways of gathering and reflecting our collective mood. But at least DeepSeek was on the right track.

Second Attempt

My revised prompt was «Using Gregory Ulmer's CATTt, develop a discourse on method that will enable a fifth estate to konsult on a super wicked problem».

Again, DeepSeek's response is methodical, defining each term — «Super wicked problems are those with time-sensitive, complex, and multi-stakeholder issues where no central authority can enforce solutions, like climate change»[17] — and moving through each part of the CATTt. It also reflects on and summarizes its work:

> Now, putting this together. The discourse on method would outline a process where the Fifth Estate uses digital platforms (electracy) to facilitate a decentralized, participatory approach to super wicked problems. Contrasting with the Fourth Estate's role in informing, the Fifth Estate actively engages in collaborative problem-solving. The analogy might be a digital agora or a swarm intelligence approach. The theory underpinning this would be electracy, collective intelligence, and maybe actor-network theory. The target is the networked public sphere, and the tale would be a case study of this method applied to a real super wicked problem.[18]

DeepSeek then produces a tale and discusses potential challenges to implementing it:

> tale: A narrative where a community uses a digital platform to crowdsource solutions to a local environmental crisis, integrating diverse perspectives, using data visualization, and iterative feedback loops to adapt strategies in real-time. Potential challenges: Ensuring inclusivity, avoiding echo chambers, managing conflict, integrating technical and local knowledge. The method must address these through design features — like algorithms that ensure diverse viewpoints are represented, facilitation by AI mediators, or structured deliberation processes. Also, considering Ulmer's emphasis on the

---

[17] Response generated by DeepSeek.
[18] Response generated by DeepSeek.

«consult» in «konsult», which he differentiates from traditional consultation by making it more immersive and integrated with digital media. So the method might involve creating a virtual environment where stakeholders can engage in simulations, role-playing, or other experiential learning to understand the problem's complexities.[19]

In the table below I summarize DeepSeek's second attempt:

CATTt	Contrast	Analogy	Theory	Target	tale
DeepSeek Second Attempt	Fourth Estate and traditional governance models	Neural networks	Theories of collective intelligence (Lévy) and actor-network theory (Latour) underpin the integration of human and non-human actors (e.g., AI, sensors). Complexity theory informs adaptive strategies, while agonistic pluralism (Mouffe) navigates conflicting stakeholder interests	Empowered citizens, grassroots organizations, and AI mediators, utilizing platforms like decentralized social networks, collaborative tools (e.g., wikis, Miro), and immersive environments (VR/AR)	A digital agora platform

Again, I am very impressed with the work. It is better than DeepSeek's response to the first prompt, yet I recognize that we are still operating within the logic of «problems-solutions», which tells me that DeepSeek lacks the necessary perspective

---

[19] Response generated by DeepSeek.

embedded in Ulmer's slogan *Problems B Us*. The literate model of self-knowledge is introspection: looking inside to see what's there and how one's mind and emotions work. The electrate way, by contrast, is what Lacan called «extimacy». Extimacy means the inside is out (in our habitus, in the atmosphere) and the outside is in (we internalize the superego figures and social structures around us). Therefore, the problems of the world exist «in us» and not just «out there». For Ulmer, we learn about ourselves by identifying the problems of the world and saying, «that's me». The understanding we get is more of a feeling than a concept; the feeling is «attunement» — a recognition that the inside mirrors the outside. Ulmer frames this work in relation to the traditions of sacred sociology. With that framework in mind, I tried again.

DeepSeek Third Attempt

For the third attempt, I wanted DeepSeek to downplay practical solutions and instead reach a spiritual understanding of our condition, one rooted in grief. I told DeepSeek that civilizational collapse was imminent: «Assuming imminent civilizational collapse as a given, use Gregory Ulmer's CATTt to develop a discourse on method that will enable a fifth estate to konsult on a super wicked problem».

DeepSeek did the best it could — «Now, considering the super wicked problem of civilizational collapse, the Fifth Estate's role would be to create a consultative process that's adaptive, inclusive, and leverages distributed knowledge»[20] — but, like many people, it avoided grief.

---

[20] Response generated by DeepSeek.

CATTt	Contrast	Analogy	Theory	Target	tale
DeepSeek Third Attempt	Linear Governance	Indigenous Praxis (stewardship systems) and Distributed Networks	Risk Society (Ulrich Beck) Cyborg Ethics (Donna Haraway) Electracy (Gregory Ulmer) Parrhesia (Foucault)	Decentralized Platforms and Grassroots Networks scale hyperlocal actions globally	The Story of Resilient Assembly

It ends suggestively: «The Fifth Estate thrives not by solving collapse but by democratizing the capacity to renegotiate survival — turning crisis into a crucible for electracy's birth».[21] Here, DeepSeek hints at grief and sacred sociology, but does not articulate it explicitly. I thus made the fourth prompt more explicit.

DeepSeek Fourth Attempt

My prompt was, «Using Gregory Ulmer's CATTt, develop a discourse on method for an electrate sacred sociology for addressing climate change». DeepSeek's response touched on sacred sociology in numerous places.

- o Contrast: Traditional sociology is empirical, maybe secular, while sacred sociology would blend spiritual values with data-driven methods.

- o Analogy: Ecosystems as an analogy for how sacred sociology can model interconnected human and environmental systems.

- o Theory: Sacred sociology might draw from theorists like Durkheim on the sacred, maybe Bruno Latour's actor-network-theory, or new materialism.

- o Target: The audience or institution to address. Probably policymakers, religious communities, environmental activists, and digital platforms. Need to consider how

---

[21] Response generated by DeepSeek.

each group interacts with climate change and how sacred sociology can engage them through electrate means like social media, virtual rituals, etc.

- o Tale: Maybe use a specific example, like the role of indigenous communities in climate activism, which combines sacredness with ecological practice. Or look at digital campaigns like #FridaysForFuture as electrate rituals.[22]

Then, as if catching itself, DeepSeek paused and added

Wait, the user mentioned «electrate sacred sociology». Electrate as per Ulmer is about the apparatus of digital media, so sacred sociology here would integrate spiritual or sacred dimensions with digital tools. So combining community rituals, digital activism, data visualization with a sacred or moral imperative.[23]

Here is the CATTt that DeepSeek generated:

CATTt	Contrast	Analogy	Theory	Target	tale
DeepSeek Fourth Attempt	Traditional sociology[24]	The Internet as a Mycelial Network	Electracy, Deep Ecology, and Ritual Studies	Institutions and Platforms for Intervention	Case study of the Green Faith movement or Extinction Rebellion's use of art and ritual in protests. Or digital memorials for extinct species as sacred electrate practices

The tale in DeepSeek's fourth response aligns with Ulmer's call to assemble electronic monuments, detailed in his 2005

---

[22] Response generated by DeepSeek.
[23] Response generated by DeepSeek.
[24] DeepSeek adds this not to the contrast: «Contrast example: While traditional sociology might analyze carbon pricing, sacred sociology

book titled *Electronic Monuments*. After this response, I felt more optimistic about DeepSeek's value as an assistant in this process. DeepSeek was discovering pathways that both Umer, and myself, had developed (I published a book chapter and article describing a proposed Monument to Lost Data to be built on the mycelial structure).[25] Of course, DeepSeek may have read our work. Yet I felt disappointed that DeepSeek had too broad an understanding of sacred sociology and I also felt that it should focus on the work of the College of Sociology, the members of which had grasped both the grammatological shift away from print and that fascism had hijacked the sacred in everyday life.[26]

DeepSeek Fifth and Final Attempt

My prompt for the fifth attempt reads: «Using Gregory Ulmer's CATTt and the sacred sociology of George Bataille and Michel Leiris, develop an electrate discourse on method for addressing climate change». DeepSeek writes

> Okay, let's tackle this query. The user wants me to develop an electrate discourse on method for addressing climate change using Gregory Ulmer's CATTt and the sacred sociology of Bataille and Leiris. Hmm, electracy [...] that's Ulmer's concept, right? Like the transition from literacy to digital media. CATTt stands for Contrast, Analogy, Theory, Target, tale. Need to remember each component.

> First, Contrast. Ulmer says to find an opposition. For climate change, maybe the contrast between utilitarian approaches (like carbon credits) and sacred/ritual practices. Bataille's ideas on excess and sacred violence versus Leiris's personal rituals. That could work.

> Then Analogy. Need a metaphor. Maybe «collective effervescence» from Durkheim, but Bataille has the notion of expenditure. So framing climate action as a ritual expenditure, a sacred act rather than just economic. That connects Bataille's sacred with Ulmer's analogy.

> Theory. Here, combine Bataille's general economy where excess is key, and Leiris's ethnography of the self.

---

would design augmented reality "mourning apps" visualizing lost ecosystems, merging data with emotional-spiritual engagement», Response generated by DeepSeek.

[25] Mauer 2006; Mauer 2008.

[26] Hollier 1988.

Apply that to climate discourse. Instead of scarcity, focus on abundance and sacrifice as sacred acts. Leiris's autobiographical approach could personalize climate narratives.

Target. The audience here is electrate society, so using digital platforms. Memes, VR, social media to create immersive experiences. Make the sacred tangible through technology. Maybe virtual rituals or augmented reality art installations.

Tale. The narrative needs to be a participatory myth. Maybe a digital pharmakon, where the poison (technology causing climate issues) is also the cure. Encourage user-generated content where people share their climate rituals, blending personal and collective.

Wait, am I mixing things correctly? Let me check Bataille's sacred sociology. He talks about transgression, excess, and the sacred as something that disrupts the profane. Leiris focused on rituals and personal identity. So integrating these into climate methods would mean creating spaces where people can engage in communal, almost ritualistic actions that transgress the usual capitalist approaches.

Also, electrate discourse requires using digital media as the primary tool. So examples like VR simulations of climate impacts, social media challenges promoting sustainable rituals, or interactive art that symbolizes sacrifice. The key is to make the abstract concept of climate change emotionally engaging through sacred experiences.[27]

---

[27] Response generated by DeepSeek. Ulmer says digital media is «not» required to practice electrate discourse.

CATTt	Contrast	Analogy	Theory	Target	tale
DeepSeek Fifth Attempt	Utilitarian techno-solutions	Climate action as collective ritual expenditure (Bataille's *dépense*)	Bataille: Climate collapse as a limit-experience (*expérience limite*), dissolving individualism into cosmic continuity. Leiris: Mythobiography — blending personal and planetary narratives Ulmer: Use digital media to construct a popcycle (public opinion) around climate sacrality — memes, games, VR ecotraumas	Electrate society — digital natives attuned to hybrid realities	A participatory myth of climate pharmakon (poison/cure)

DeepSeek summarizes its work below:

Electrate Ritual Prototype: «The Carbon Potlatch»

1. Contrast: Replace carbon offsets with gift economies (e.g., plant trees as offerings, not transactions).
2. Analogy: Hack corporate platforms — flood Instagram with #SacredMethane leaks (aestheticize pollution as grotesque ritual).
3. Theory: Deploy Bataille's Base Materialism — celebrate decay (compost TikTok ASMR) as anti-capitalist sacrament.

4. Target: Gamify IPCC reports into Mythic Quests (e.g., «Solve the Permafrost Labyrinth» in Fortnite).
5. Tale: Crowdsource a Global Mourning Mask (3D-printed from ocean plastic), worn in Zoom climate summits.

Conclusion: Electrate Sacrality as Method

> By fusing Ulmer's digital pragmatism with Bataille/Leiris' sacred transgression, climate action becomes a liturgical practice — transmuting despair into collective jouissance, hacking technocapitalist tools to ritualize survival. The method is the ritual: the medium (TikTok, VR, DAOs) is the temple.[28]

DeepSeek then asked, «Would you like to expand on any specific element (e.g., AR rituals, Bataille's sacrifice)?»[29] I responded, «Expand on Bataille's sacrifice».

Sacrifice in an Electrate Sacred Sociology of Climate Change

Georges Bataille's concept of «sacrifice» — central to his theory of *The Accursed Share* — offers a radical framework for rethinking climate action as a «ritual expenditure of energy», rather than just rational conservation. In an electrate sacred sociology, Bataille's ideas help us «ritualize waste, excess, and destruction» in ways that bind communities together in the face of ecological collapse.

1. Bataille's Core Idea: The Necessity of Expenditure

Bataille argues that all societies produce «excess energy» (the «accursed share») that cannot be productively reinvested. This surplus must be «squandered» in glorious, non-utilitarian ways, through:

- o Sacrifice (destruction of wealth/goods to honor the sacred)
- o Potlatch (competitive gift-giving that destroys value)
- o Festival & Orgy (ecstatic consumption beyond necessity)

For Bataille, «modern capitalism fails because it hoards rather than sacrifices», leading to catastrophic imbalances (e.g., climate change as hoarded carbon excess).

2. Sacrifice as Climate Ritual

To apply Bataille to climate change, we must «reintroduce sacred waste» into ecological practice:

A. *Symbolic Carbon Sacrifices*
- o «Burn Your SUV (Virtually) »: A digital ritual where users 'sacrifice' their carbon footprints in an AR bonfire.

---

[28] Response generated by DeepSeek.
[29] Response generated by DeepSeek.

○   «Crypto Burn Events»: Like the Maya burning precious goods, we destroy Bitcoin (a high-energy commodity) as atonement.

B. *The Great Climate Potlatch*
   ○   «Corporate Waste Wars»: Companies compete to *destroy* their own excess (e.g., Amazon 'sacrifices' returns instead of dumping them).
   ○   «Debt Forgiveness as Sacrifice»: Nations annihilate carbon debt in a global potlatch.

C. *The Feast of Extinction*
   ○   «Last Meals for Dying Species»: Communal dinners where participants consume dishes mimicking extinct animals, then «destroy the leftovers» (cf. Bataille's *The Solar Anus*).

3. Electrate Sacrifice: Digital Platforms for Sacred Waste

Bataille's sacrifice must be «translated into electrate rituals»:

A. Twitch Sacrifice Streams
   ○   Viewers donate to 'kill' digital avatars of oil executives or melting glaciers — «destruction as catharsis».
   ○   «Click to Burn»: A collective button-press that «sacrifices» a virtual forest, mirroring real conservation.

B. AI-Mediated Scapegoats
   ○   ChatGPT generates «climate sin confessions» (e.g., «I flew to Bali») and offers «AI-prescribed penance» (e.g., «Donate to coral reefs»).
   ○   «Deepfake Sacrificial Kings»: Algorithmically generated figures (like Bataille's *Acephalic* headless god) who 'die' for our carbon sins.

C. NFT Sacrifice Pits
   ○   «Burnable NFTs»: Artists create tokens representing endangered species; collectors must «destroy them» to «save» the real creature.

4. Why This Works: The Sacred in the Profane

Bataille insists that «only through sacrificial loss» can we reconnect with the sacred». Climate discourse is trapped in «moralizing guilt» («reduce, reuse, recycle») — but guilt is individual, while «sacrifice is collective».

   ○   Example: Iceland's *Funeral for a Glacier* (2019) was Bataillean — a public mourning for lost ice, «useless yet necessary».
   ○   Contrast: Carbon taxes are transactional; «carbon sacrifices» are *liturgical*.
   ○   Risks & Counterarguments

○   Does ritualizing waste normalize destruction?

○   Bataille: No — sacrifice is *conscious* waste, unlike capitalist waste (e.g., planned obsolescence).

○   Isn't this just slacktivism?

○   Electrate rituals must tie to material acts (e.g., NFT burns fund rewilding).

Conclusion: The Electrate Altar

Bataille's sacrifice offers a way to «transmute climate despair into sacred theater». By designing «digital rituals of expenditure», we turn the Anthropocene's waste back into «communion».

«To save the world, we must first learn to lose it».

Many of the elements in DeepSeek's fifth response — and its supplemental expansion on Bataille's sacrifice — are bold, transgressive, and even profound. But I question some of its outputs.

1. «Nations annihilate carbon debt in a global potlatch».[30] I wondered if this action would hurt developing countries and block efforts to slow climate destruction. But I found an article supporting the idea: «The starting point for such treaty could be a mutual debt cancellation, developed countries' carbon debts offsetting developing countries' conventional monetary debt, leaving the dispute about historical responsibility behind».[31]

2. «Last Meals for Dying Species: Communal dinners where participants consume dishes mimicking extinct animals, then 'destroy the leftovers' (cf. Bataille's *The Solar Anus*)».[32] I am confused about who is doing what. Are participants mimicking extinct animals? Or does the food look like extinct animals?

3. «Viewers donate to 'kill' digital avatars of oil executives or melting glaciers — "destruction as catharsis"».[33] The sentence structure here suggests that we kill digital avatars of melting glaciers. Are glaciers alive? If they are melting, aren't they already dying? Then I noticed further down in its output that DeepSeek cites Iceland's 2019 Funeral for a Glacier, a real event.[34] Killing digital

---

[30] Response generated by DeepSeek.
[31] Kunnas 2014.
[32] Response generated by DeepSeek.
[33] Response generated by DeepSeek.
[34] Eno 1973.

avatars of oil executives is a bad idea for a number of reasons, including that it could lead to charges of murder and terrorism were someone to act on it in real life.

4. «"Click to Burn": A collective button-press that 'sacrifices' a virtual forest, mirroring real conservation».[35] How is sacrificing a virtual forest mirroring real conservation? Isn't it mirroring real destruction?

The slogan that DeepSeek provided at the end of its output — «To save the world, we must first learn to lose it»[36] — struck me as original and profound. I did a Google search to see if it was a quote but turned up nothing. DeepSeek appears to have 'thought' of it on its own.

Conclusion

How do we assess DeepSeek's contribution to this effort to develop an electrate discourse on method? I believe that Brian Eno's statement, quoted at the beginning of this essay, helps me make sense of it. DeepSeek enables me to «replace the element of skill — considered necessary in [developing a discourse on method] — with the element of judgment».[37] My decades of experience in producing discourses on method and in working with the CATTt model in relation to a variety of problems has taught me to distinguish between low value and high value outputs. Further, it has enabled me to engineer my prompts to elicit better and better responses. I could take any of the high value elements of DeepSeek's outputs and enter them back into DeepSeek — as I did with the Bataille prompt — and elicit additional outputs. The results are fascinating, if at times difficult to understand (sometimes because of ambiguous sentence construction). These experiments, conducted in under an hour (though spread over a couple of days), reveal that Eno was right. Skill can be replaced with judgment. But, I would argue, judgment cannot be gained without experience, and experience is hard to gain without learning to develop some skill.

---

[35] Response generated by DeepSeek.
[36] Response generated by DeepSeek.
[37] Eno 1973.

Bibliography

Eno, Brian. 1973. (TV-MASTER) «Iceland Holds Funeral for First Glacier Lost to Climate Change». *The Guardian*, 18 August 2019. ‹https://youtu.be/uj664-SkU3c?si=8bHOG dP-6xScgkeb›

Havelock, Eric A. 1986. *The Muse Learns to Write: Reflections on Orality and Literacy from Antiquity to the Present.* New Haven, Yale University Press.

Hollier, Denis, ed. 1988. *The College of Sociology (1937-39).* Minneapolis, University of Minnesota Press.

Kerner, Sean Michael. 2025. «DeepSeek Explained: Everything You Need to Know». *TechTarget* 26 August 2025. ‹https://www.techtarget.com/whatis/feature/DeepSeek-explained-Everything-you-need-to-know›

Kunnas, Jan, Eoin McLaughlin, Nick Hanley, David Greasley, Les Oxley and Paul Warde. 2014. «Counting Carbon: Historic Emissions from Fossil Fuels, Long-Run Measures of Sustainable Development and Carbon Debt». *Scandinavian Economic History Review* 62.3: 243-65.

Mauer, Barry. 2006. «18. Proposal for a Monument to Lost Data». In *Studies in Writing* 17. *Writing and Digital Media.* Van Waes, Luuk, Mariëlle Leijten and Christine M. Neuwirth, eds. Amsterdam, Elsevier Press. 287-309.

Mauer, Barry. 2008. «Lost Data, 2». *Rhizomes* 18. ‹http://www.rhizomes.net/issue18/mauer/index.html›

Ulmer, Gregory. 1994. *Heuretics: The Logic of Invention.* Baltimore, The Johns Hopkins University Press. 15.

Ulmer, Gregory. 2003. *Internet invention: From Literacy to Electracy.* Boston, Addison-Wesley Longman. 24-5.

Ulmer, Gregory. 2005a. «The Learning Screen from Networked Book». ‹https://www.academia.edu/37590082/The_Lear ning_Screen_From_Networked_Book›

Ulmer, Gregory. 2005b. *Electronic Monuments.* Minneapolis, University of Minnesota Press.

Wikipedia. 2025. «Wicked problem». ‹https://en.wikipedia.org/wiki/Wicked_problem›

Cartografía de un cruce en desarrollo: perspectivas e implicaciones de la IA en la traducción literaria

Guillem Molla

La traducción automática y sus siglas: de la TA a los GPT

Discutir en un artículo impreso la aplicación de la inteligencia artificial (IA) a la traducción literaria implica asumir, desde el inicio, su inevitable obsolescencia. Cuestiones como la falta de rigor lingüístico, de creatividad o de contextualización pragmática, junto a consideraciones de orden estructural o sistémico como el sesgo algorítmico, la opacidad de los sistemas, el coste medioambiental o los dilemas éticos y legales, son evaluadas de forma constante mediante propuestas técnicas o aportaciones teóricas orientadas, al menos en parte, a su resolución.

La implicación de la IA en la traducción se refleja en su propia definición general de diccionario en inglés.[1] En los estudios de traducción, la traducción automática (TA, o MT, *machine translation*) se define esencialmente como la conversión automática de un texto de una lengua natural a otra mediante una máquina, incluso cuando su funcionamiento depende del trabajo humano antes, durante o después del proceso.[2] El nuevo paradigma, consolidado por los avances que combinan el aprendizaje profundo (DL, *deep learning*) y el procesamiento

---

[1] Según el *New Oxford American Dictionary* (3.ª ed.), la IA es «the theory and development of computer systems able to perform tasks that normally require human intelligence, such as visual perception, speech recognition, decision-making, and translation between languages». El *Cambridge Dictionary* destaca la capacidad de interpretar y producir lenguaje, mientras que *Merriam-Webster*, de forma más general, alude a la imitación del comportamiento humano inteligente. Aunque muchos diccionarios normativos de otras lenguas — como el *DLE* (RAE), el *DLP* (Academia das Ciências de Lisboa) o el *DIEC* (IEC) en el ámbito peninsular — no mencionan la traducción, diversas fuentes técnicas y académicas la consideran una de las aplicaciones más relevantes de la IA.

[2] Kenny 2019, 305-306.

del lenguaje natural (PLN, o NLP, *natural language processing*), supone un reto para la práctica docente de la traducción, mientras que en la práctica profesional en áreas técnicas, científicas y administrativas constituye ya una realidad cada vez más asentada.[3] También lo es en el campo de la traducción literaria, «the most demanding type of translation»,[4] según Landers, pues no solo requiere fidelidad al contenido, sino también una gran sensibilidad estética y un dominio profundo tanto de la lengua de partida como de la lengua meta. La buena comprensión del estilo literario, sigue Landers, «can make the difference between a lively, highly readable translation and a stilted, rigid, and *artificial* rendering that strips the original of its artistic and aesthetic essence, even its very soul».[5]

El interés por las implicaciones de la IA en la traducción literaria se ha incrementado de manera considerable en los últimos años. Tras la (r)evolución técnica de los antiguos sistemas basados en reglas (RBMT, *rule-based machine translation*) y los modelos estadísticos (SMT, *statistical machine translations*) a partir de los años noventa, el desarrollo de la traducción automática neuronal (TAN, o NMT, *neural machine translation*) desde 2014 y, sobre todo, su salto cualitativo y democratización funcional a finales de 2022, ha generado un impacto mediático, educativo y profesional rico en reflexiones y estudios de caso. Al lado de la TAN, que es un enfoque que emplea grandes corpus paralelos para predecir secuencias lingüísticas completas y que en el sector comercial sigue siendo «the dominant approach today»,[6] los grandes modelos de lenguaje (LLM, *large language models*), entre los que se incluyen los transformadores generativos preentrenados (GPT, *generative pretrained transformers*), están ganando terreno en escenarios multilingües.[7] Así, se puede observar que mientras algunas investigaciones exploran la mejora de resultados mediante la ingeniería de instrucciones (*prompt engineering*), otras muchas siguen la línea de comparar las traducciones generadas por las diversas herramientas de IA y de estas con las humanas, a menudo con el objetivo de contribuir al desarrollo futuro de este

---

[3] Massey 2023; Way 2023; Robinson 2023; L. Wang 2023; Rico Pérez 2024; Nagy 2025; Zhang 2025.
[4] Landers 2001, 7.
[5] Landers 2001, 7.
[6] Koehn 2020, 293.
[7] Moorkens 2025, 1-24.

campo en ebullición. Este artículo ofrece una cartografía analítica de investigaciones recientes sobre los avances tecnológicos en traducción automática y sus efectos en el ámbito literario, trazada desde una perspectiva crítica, creativa y ética. Dado el crecimiento exponencial de estudios empíricos, el análisis se acota a una selección representativa de casos.

Más allá de la eficiencia: el traductor humano frente al espejo de la posedición

Si hasta hace poco tiempo los resultados obtenidos por los sistemas de traducción automática — particularmente cuando no hay intervención humana — se consideraban decididamente problemáticos en términos de precisión y preservación de la creatividad,[8] investigaciones más recientes indican que, gracias a los avances en la TAN y a la implementación de LLM entrenados con corpus masivos — como los que emplean ChatGPT, Gemini (antes Bard), Microsoft Copilot o Baidu —, así como en traductores especializados como DeepL o Google Translate (GT), los sistemas actuales son capaces de producir traducciones menos literales, con mejoras perceptibles en fluidez, manejo del lenguaje figurado y mayor sensibilidad al contexto. Los enfoques, sin embargo, varían según el entorno cultural y lingüístico. En los países anglófonos, la investigación se centra en el perfeccionamiento de algoritmos, el tratamiento contextual y los métodos de evaluación, mientras que en China se prioriza el desarrollo de aplicaciones móviles y plataformas en línea que integran la IA en la práctica traductora.[9]

Los resultados de algunos estudios publicados en los dos últimos años sobre traducción automática de poesía entre chino e inglés, tomando como ejemplo una combinación especialmente compleja por la distancia tipológica y la carga literaria de los textos, coinciden en destacar estos avances más allá de la velocidad y la practicidad. Hu y Xiaoqian señalan que DeepL alcanza una tasa de éxito superior al 80% en precisión y fluidez al traducir dos obras de Shakespeare, y muestra cierta creatividad mediante recursos como la adición, la explicitación o el cambio de perspectiva.[10] De forma similar, Gao

---

[8] Guerberof Arenas 2014; Gaspari 2016; Sánchez-Torrón 2016.
[9] Pan 2024, 29.
[10] Hu y Xiaoqian 2023.

concluye que ChatGPT-3.5, a pesar de no estar concebido originalmente para la traducción, incluso supera a DeepL y GT en fidelidad, estilo lingüístico y calidad global al traducir poesía clásica china, singularmente cuando se emplea un *prompt* que solicita conservar el ritmo y la rima, haciendo que su estilo sea «less robotic».[11] De manera complementaria, el método EAPMT (*explanation-assisted poetry machine translation*), desarrollado por Wang, utiliza explicaciones monolingües como guía y, a su parecer, mejora el estilo poético y la segmentación de versos en traducciones de poesía inglesa obtenidas por ChatGPT.[12] No obstante, todos los estudios subrayan también limitaciones de peso, que van desde errores léxicos y sintácticos[13] hasta problemas de coherencia, alucinaciones y una comprensión insuficiente del trasfondo cultural o del sentido original.[14] En efecto, tanto estos estudios como los centrados en otras lenguas concuerdan en afirmar que la posedición (PE) — entendida como la supervisión humana tras una propuesta generada automáticamente — sigue siendo necesaria para garantizar la calidad global de la traducción.

Una de las principales cuestiones que se plantean es cómo aprovechar estas herramientas sin caer en un «discurso solucionista»,[15] concibiéndolas como una oportunidad y un recurso en beneficio de las personas traductoras. Conforme a esta idea, y sin ignorar las reticencias de muchos profesionales ante la expansión de las nuevas tecnologías, se ha propuesto que estos modelos sirvan para acompañar, más que sustituir, el trabajo creativo del traductor, favoreciendo una interacción flexible que mantenga su protagonismo en el proceso.[16] Esta estrategia, en su formulación más favorable, permitiría maximizar la eficiencia sin comprometer la integridad literaria. Al hilo de lo dicho, el estudio comparativo llevado a cabo por Castaldo es ilustrativo. A través de una herramienta personalizada, llamada UniOr-PET, y con la colaboración de traductores profesionales, se evalúa la viabilidad de la PE de traducciones literarias generadas por LLM de fragmentos de una novela de Margaret Atwood al italiano, analizando el tiempo de edición,

---

[11] Gao 2024, 5.
[12] Wang 2024.
[13] Hu y Xiaoqian 2023.
[14] Wang 2024; Gao 2024.
[15] Declercq y Van Egdom 2023, 50.
[16] Way, 2023.

la calidad y la creatividad de GPT-4, GPT-3.5 y un modelo Mistral-7B adaptado para literatura. Su conclusión es que, en el marco de su estudio, «creativity does not present a significant difference between human translation and post-edited LLM translations»,[17] lo que sugiere que los modelos ajustados mediante *fine-tuning* con corpus literarios, incluso si son pequeños, pueden lograr un buen equilibrio entre calidad y eficiencia en la traducción.[18] Por su parte, el estudio empírico de Li demuestra que la PE de textos creativos con TAN no solo incrementa la eficiencia, sino que también puede dar lugar a resultados más aceptables que una traducción hecha desde cero, aunque los estudiantes de traducción participantes en el estudio señalan, a su vez, que se pierde parte del estilo y la naturalidad del texto original. Se advierte, además, que deben considerarse ciertos efectos secundarios, como el impacto en la capacidad crítica de los usuarios y las cuestiones éticas vinculadas al plagio, pero sostienen que, en lugar de evitar su uso por estos motivos, «we need to explore and learn how to use it ethically, for working with NMT and other AI technologies is inevitable».[19]

Otras voces plantean una lectura más escéptica. Sin duda, la expansión de estos sistemas suscita preguntas en torno a la pérdida del rol autoral del traductor humano, asociado al riesgo de reconvertirlo en corrector técnico, en *revisor* o *poseditor* subordinado al primer discurso generado por la máquina bajo el efecto de la imprimación (*priming*); por lo tanto, sin mucho margen de maniobra y, por consiguiente, más cercano a la reparación que al proceso creativo.[20] Según Nagy, algunas editoriales ya utilizan DeepL para traducir libros y luego revisarlos «[t]o make it seem more "human"»,[21] dando lugar a textos corregidos en exceso que empobrecen su estilo. Desde una perspectiva cultural, Nagy sostiene que la traducción literaria constituye una forma de experiencia no solo lingüística, sino también vital, lo cual equivale a «translating something lived into something written [...] moving between an experience in one language into an experience in another language,

---

[17] Castaldo 2025, 7.
[18] Castaldo 2025, 7.
[19] Li 2025, 10.
[20] Moorkens 2018; Kenny 2020; Guerberof-Arenas y Toral 2022; Li 2023; Mirghaderi 2023; H. Wang 2023; Winters 2024; Rico Pérez 2025.
[21] Nagy 2025, 656.

plus adding some experience of the translator».[22] Las herramientas basadas en IA pueden cumplir funciones comunicativas básicas, pero no pueden sustituir «our humanity and the mystery of mankind».[23] En el estudio de Zhang, se alerta sobre estos riesgos profesionales, ya que «naive translation companies might eliminate human translators due to misjudgments based on MQM [Multidimensional Quality Metrics] or SQM [Scalar Quality Metrics] evaluations by inexperienced practitioners».[24] También frente al entusiasmo automatizador, partiendo de la premisa de que la profesión del traductor se encuentra en una encrucijada, Katan propone reforzar el valor añadido del traductor humano frente a la máquina, «actively demonstrating what machines can't do by becoming a *homo narrans* translator»[25]; a saber, mediante una práctica traductora que no solo reproduzca el contenido del original, sino que lo reconfigure narrativamente en función de las necesidades del nuevo público lector.[26]

En la misma línea, y ante los desafíos éticos y profesionales que plantea la IA en la formación académica de los futuros traductores, Spoturno recuerda «la importancia de considerar la obra en su totalidad y de valorar la traducción literaria como una actividad creativa y humana», algo que los sistemas de TAN no pueden ofrecer al generar «productos fragmentarios y homogenizados».[27] La figura del oxímoron sirve a Leonardi para subrayar la contradicción entre el lenguaje poético —cargado de ambigüedad y sensibilidad estética — y el funcionamiento literal de DeepL, GT y Yandex. A partir de una selección de poemas de Giacomo Leopardi, con los que evalúa las limitaciones de estos sistemas, concluye que «AI is unsuited for Italian to English poetry translation without human intervention»,[28] pues no logran interpretar adecuadamente ni el significado ni la referencialidad poética del autor. Por otro lado, desde una postura más pragmática y en relación con su estudio sobre una tragedia en alemán de Heinrich Kruse, Portelli

---

[22] Nagy 2025, 656.
[23] Nagy 2025, 656-57.
[24] Zhang 2025, 10970.
[25] Katan 2023, 87.
[26] Katan 2023, 87.
[27] Spoturno 2024, 16.
[28] Leonardi 2024, 647-48.

reconoce las limitaciones de la traducción automática en textos estilísticamente complejos. Sin embargo, aunque «[t]ranslation technology is not a tool that is often looked upon favourably by literary scholars»,[29] defiende su utilidad en la investigación literaria comparada, como es el caso, al permitir superar barreras lingüísticas que históricamente han limitado el acceso a ciertas obras no canónicas o nunca traducidas.

Fijémonos ahora en cómo responden estos modelos cuando se trata de lenguas no hegemónicas en contextos occidentales, como el catalán; es decir, lenguas no globales que, a pesar de su gran vitalidad sociolingüística y de los avances recientes en la compilación de corpus, siguen contando con recursos comparativamente más limitados que lenguas como el inglés. Tras unos primeros resultados prometedores con modelos estadísticos aplicados a la traducción de lenguas románicas o cercanas — español y catalán —,[30] se ha destacado la mayor eficiencia productiva de los modelos neuronales — en palabras de los autores, «the first experiment in the literature in which a fragment of a novel is translated automatically and then post-edited by professional translators»,[31] lo que situaría al catalán como la primera lengua meta documentada en este tipo de estudios de caso literario —, si bien se ha constatado también una preferencia generalizada por el estilo y la creatividad de la traducción humana.[32] Sobre esta base, los resultados de Guerberof-Arenas y Toral en tres modalidades de traducción de un relato de ficción del inglés al catalán revelan que, entre los participantes catalanohablantes encuestados, «HT [human translation] scores higher for creativity than MTPE [machine translation post-editing] does».[33] Además, «HT scores higher in narrative engagement and translation reception and is slightly lower than MTPE in enjoyment»,[34] dejando claro que «professional translators, by providing solutions that are both novel and acceptable, add the creativity factor that MT [machine translation] is lacking at present».[35] Posteriormente, a partir del análisis comparativo en la traducción de

---

[29] Portelli 2024, 33.
[30] Toral y Way 2015.
[31] Toral 2018, 10.
[32] Moorkens 2018.
[33] Guerberof-Arenas y Toral 2020, 276.
[34] Guerberof-Arenas y Toral 2020, 276.
[35] Guerberof-Arenas y Toral 2020, 277.

un relato de ciencia ficción del inglés al catalán y al neerlandés, Guerberof-Arenas y Toral añaden que, según los nuevos datos obtenidos, «using MT (by means of PE [post-edition]) hinders the effectiveness of the translation process, because the translator becomes the evaluator and not the creator, and therefore the mechanisms and phases of creativity are not set into motion».[36]

En otras palabras, aun con los avances tecnológicos, «a substantial gap remains between LLM and human quality in literary translation»[37] en aspectos clave como el tono, la emoción y el uso de recursos estilísticos. Las métricas de evaluación actuales privilegian la exactitud mecánica por delante de la creatividad, con lo cual sería recomendable avanzar en métricas que evalúen el estilo y la terminología para reflejar mejor la calidad literaria.[38]

Lo que la IA (todavía) no entiende: sesgos, límites y propuestas

Asimismo, y en estrecha relación con las elecciones estilísticas, tampoco existe una métrica universal para evaluar la dimensión ética de los sistemas algorítmicos aplicados a la traducción, ya que varían según intereses corporativos y criterios de eficiencia.[39] Según recomienda Mirghaderi, si los conjuntos de datos utilizados para entrenar algoritmos se hacen públicos, se evalúan por entidades supervisoras y se comparten entre desarrolladores, «biases will be exposed, and these datasets can be improved over time».[40] A este problema de interpretabilidad podrían dar respuesta nuevos enfoques como la inteligencia artificial explicable (XAI, *explainable artificial intelligence*), cuyo objetivo es hacer comprensibles la toma de decisiones algorítmicas). Cabe tener en cuenta que estos sistemas y modelos, lejos de ser neutros, pueden producir sesgos raciales, políticos y de género. Tal como argumentan Monzó-Nebot y Tasa-Fuster, al estar entrenados con grandes volúmenes de datos históricos y programados para priorizar los patrones más frecuentes, «issues result from the biases present in soci-

---

[36] Guerberof-Arenas y Toral 2022, 207.
[37] Zhang 2025, 10970.
[38] Zhang 2025, 10970.
[39] Moorkens 2023; Li 2023.
[40] Mirghaderi 2023, 841.

ety (its texts), the low representativity of most advanced discourses in the overall corpora, and a conception of the technology whereby priority is granted to what is already prioritized».[41]

En este sentido, los estudios recogidos en la tercera parte del volumen *Gendered Technology in Translation and Interpreting*[42] analizan de qué manera la traducción automática neuronal y los grandes modelos de lenguaje no solo pueden perpetuar, sino incluso amplificar los sesgos de género. Así, en el análisis sobre el tratamiento de nombres de profesiones en textos institucionales traducidos del inglés al alemán, las herramientas estudiadas — incluida la base terminológica de la Unión Europea (IATE) — tienden a reproducir una masculinización sistemática de cargos neutros o ambiguos.[43] En la comparación de traducciones al gallego realizadas tanto por sistemas de TAN como por estudiantes traductores, se observa una asignación de género estereotipada — por ejemplo, «nurse» como «enfermeira« cuando en la trama general de la obra se está hablando de personajes masculinos — que evidencia la necesidad de estrategias pedagógicas para corregir estos patrones en máquinas y, también, en los propios traductores humanos en formación.[44] Las herramientas de GT y ChatGPT, a pesar de intentar evitar la reproducción de estereotipos — el primero generando dos traducciones alternativas y el segundo, tras un segundo prompt, con opciones dobles como «el/la abogado/a» —, siguen limitadas por una perspectiva que excluye las identidades no binarias.[45] De hecho, incluso formulando instrucciones precisas, «GPT models perpetuate biases even when explicitly prompted to provide alternative translation».[46]

Estos trabajos ponen de manifiesto que no basta con incorporar un criterio ético y terminológico a posteriori, sino que este debe estar presente desde el diseño mismo de los modelos tecnológicos, a fin de evitar que las nuevas herramientas lingüísticas, en lugar de corregir problemas estructurales, los repliquen con mayor eficiencia. Abundando en esta idea, Ghosh

---

[41] Monzó-Nebot y Tasa-Fuster 2024, 9.
[42] Monzó-Nebot y Tasa-Fuster 2024.
[43] Đorđević, 2024.
[44] García González 2024.
[45] Rico Pérez y Pleguezuelos 2024, 261.
[46] Vanmassenhove 2024, 244.

y Chatterjee, tras analizar la tendencia a malinterpretar el género en las traducciones entre el inglés y las lenguas consideradas «low-resource» como consecuencia del colonialismo, proponen estrategias sociotécnicas que vayan más allá del anglocentrismo, dado que «each language is unique, and they should not collectively be homogenized to the high(est) resourced English».[47] Implica, pues, un trabajo interdisciplinario más ambicioso, donde se involucre activamente a las comunidades lingüísticas locales en la recolección de corpus, entrenamiento y validación de los modelos en aras de promover también la sostenibilidad lingüística y la equidad cultural.[48]

De forma paralela, la traducción automática de los elementos culturales específicos (ECE, o CSI, *culture-specific items*) para las lenguas y contextos no occidentales es también motivo de atención. Opaluwah, partiendo del modelo de estrategias traductoras para ECE de Aixelá (1996), observa que, en el caso de la poesía de Wole Soyinka traducida del inglés al francés, los motores generales tienden a recurrir mayoritariamente a la repetición, mientras que los sistemas entrenados específicamente para un corpus autoral o cultural muestran una mayor diversidad de procedimientos, como la traducción lingüística, la universalización y la sinonimia. En otro estudio, Manapbayeva, comparando la traducción humana y de ChatGPT-4 de un poema en kazajo de Abay Qunanbayuli, observan que «[a]ll in all, AI follows the foreignization principle instead of domestication»,[49] cumpliendo así dos propósitos distintos: mientras que la traducción humana aporta profundidad literaria y emocional, la de la IA se caracteriza por su claridad y sencillez.[50] Asimismo, en la comparación de Sheng y Yankun entre traducciones al inglés y francés de textos chinos de contenido político e ideológico, se concluye que, a nivel semántico, «there are ideological differences between the human translation and neural machine translation».[51] Si bien es cierto que la tendencia a la literalidad de la traducción automática neuronal puede aparentemente ofrecer mayor objeti-

---

[47] Ghosh y Chatterjee 2024, 284.
[48] Ghosh y Chatterjee 2024, 281-84.
[49] Manapbayeva 2024, 529.
[50] Manapbayeva 2024, 530.
[51] Sheng y Yankun 2023, 495.

vidad, la subjetividad humana permite captar y adaptar connotaciones que estos sistemas aún no comprenden.[52] A la luz de las limitaciones de la TAN en la transferencia del sentido cultural, y a través del estudio de una novela en urdu de Khadija Mastoor, Naeem propone un marco de colaboración híbrida, de modo que «[b]y combining human insight with technological innovation, we can navigate the complexities of cultural adaptation and achieve meaningful and lasting connections across cultures».[53]

Ahora bien, ante un panorama digital cada vez más complejo, algunos de los estudios que abordan la traducción literaria del árabe expresan reservas respecto a la capacidad de la traducción automática para adaptar los ECE. Su condición de lengua global no impide que, en el ámbito tecnológico, siga enfrentando retos en el tratamiento de su ambigüedad lingüística, complejidad morfológica, ausencia de vocalización en la escritura y amplia variación dialectal.[54] En relación con este punto, los resultados del estudio de Al Rousan, centrado en la traducción al inglés de una novela de Tayeb Salih, revelan que, pese a su fluidez, las versiones generadas por GPT-4 son significativamente menos precisas (77,9 %) que las humanas (94,5 %), presentando dificultades para identificar correctamente a los agentes del discurso — lo cual se traduce en errores en los pronombres de los personajes — y, también, en la traslación de términos culturales. Por eso, aunque se advierte sobre la posible normalización de traducciones de baja calidad si la IA llega a sustituir gran parte del trabajo humano, concluyen que «the value of a good human translator would remain unchanged».[55] De igual modo, Ouldelhaj y Benmakhlouf, al comparar traducciones humanas con las automáticas de ChatGPT-3.5 de novelas y cuentos entre español y árabe, sostienen que «las habilidades humanas en este campo son prácticamente insustituibles».[56] A su juicio, la IA adolece de una «insalvable incompetencia discursiva general»,[57] ya que traduce de forma fragmentaria, sin captar el hilo narrativo ni el

---

[52] Sheng y Yankun 2023, 495.
[53] Naeem 2025, 239.
[54] Aldawsari 2025.
[55] Al Rousan 2025, 16.
[56] Ouldelhaj y Benmakhlouf 2025, 2.
[57] Ouldelhaj y Benmakhlouf 2025, 16.

trasfondo cultural que el traductor humano sí puede reconstruir gracias a su conciencia estética y su capacidad de restituir el sentido global de la obra. En línea con estas evaluaciones, Sharar confirma los problemas de literalidad, pérdida de carga emocional y falta de cohesión textual en las traducciones de *O Night*, de Khalil Gibran, realizadas por ChatGPT. Finalmente, Abdelhalim completa el círculo al hacer notar que, si bien los estudiantes de traducción literaria consultados muestran una preferencia por ChatGPT frente a GT, ambos sistemas requieren igualmente la intervención humana en la posedición «to guarantee the precise interpretation of nuanced meanings and stylistic consistency and improve translation quality».[58]

Arte, mediación y algoritmo: a modo de cierre

Hace poco más de una década, la traducción literaria era considerada el último bastión de la traducción humana.[59] Hoy, tras la popularización de la IA y su irrupción en todo tipo de áreas creativas, también este campo experimenta una evolución imparable que debe convivir con el hecho de que «[i]n its highest form, translation isn't a job but an art».[60] Sin un conocimiento profundo del oficio, el fácil acceso a estas herramientas no garantiza resultados fiables, pues «with translation, you often only get what you pay for».[61] A su vez, la capacidad para recrear formas de extrañamiento que desautomatizan la lengua, o de aludir intertextualmente de manera no probabilística, constituye, además del placer traductor, una experiencia estética para el lector, que puede disfrutar de distintas versiones en una misma lengua y apreciar su impronta histórica, dialectal, estilística o ideológica. Aunque toda traducción esté destinada a quedar desfasada, su valor reside «in their expression of how we spoke and thought and wrote in our own time».[62]

Según H. Wang, la traducción literaria responde también a un *habitus* creativo forjado a lo largo de la trayectoria del traductor. Por ello, si la traducción automática aspira a ir más allá de los productos culturales de consumo rápido y asumir el papel del traductor humano, deberá desarrollar un algoritmo

---

[58] Abdelhalim 2025, 16.
[59] Toral y Way 2014, 174.
[60] Baron 2023, 171.
[61] Baron 2023, 275.
[62] Landers 2001, 12.

comparable.[63] A esto se suma que las traducciones humanas se distinguen no solo por su precisión, sino también por una mayor empatía a la hora de transmitir matices culturales.[64] Entonces, si la posedición no siempre es suficiente ante desafíos artísticos o sesgos estructurales,[65] cabe preguntarse si es posible trasladar con fidelidad lenguas históricamente marginadas o estigmatizadas por la lógica geopolítica — como ocurre dentro de la órbita de literaturas dominantes, como el inglés, el chino o el español — respetando la riqueza estética e ideológica de sus tradiciones literarias. Frente a la visión tecnopesimista de autores como Heidegger, los filósofos contemporáneos suelen coincidir en que la tecnología no es buena ni mala en sí misma, pero tampoco neutral, ya que su sentido se define en la interacción con los seres humanos y los usos que se le dan.[66] Así, el empleo de estas herramientas, según quién las diseñe o controle, puede alimentar un tecnonacionalismo asimilacionista que refuerce, aún más, identidades hegemónicas y sus estructuras de poder o, por el contrario, convertirse en una oportunidad para preservar y proyectar internacionalmente voces minorizadas — sin olvidar las lenguas indígenas en peligro de desaparición — más allá de criterios de rentabilidad.

También desde un punto de vista posthumanista que cuestione el antropocentrismo tradicional y promueva una atención más abierta con el entorno, las nuevas tecnologías de la traducción deberían orientarse en una dirección compatible, incluso de manera programática, con la interpretación más integradora y *mundialista* de la utopía goetheana. Como sugiere Van Egdom, es esencial que la aplicación literaria de las nuevas tecnologías de la traducción respete el ideal de la *Weltliteratur*, promoviendo la circulación global de obras, en especial de lenguas periféricas, y fomentando la empatía intercultural, en contraste con enfoques centralizados o imperialistas. Pero, de nuevo, como él mismo advierte, ese acceso ampliado corre el riesgo de homogeneizar y más bien reforzar jerarquías lingüísticas si se prescinde de la mediación crítica del traductor

---

[63] H. Wang 2023, 478.
[64] Gough 2024, 412.
[65] Spoturno 2024; Ghosh y Chatterjee 2024.
[66] Moorkens 2024, 220-21.

humano.[67] Aunque el modo literal de los sistemas de traducción automática actuales pueda parecer ideológicamente imparcial, sigue siendo incapaz, por ejemplo, de insertar anotaciones, adaptar estructuras o interpretar matices ideológicos cuando la obra lo requiere.[68] Los paratextos generados por el traductor humano — ya sean peritextuales, como prólogos, notas, glosarios, o epitextuales, como artículos, estudios, correspondencia con el autor — suplen de forma consciente lo que los modelos aún no articulan, sobre todo cuando se trata de hacer visible y comprensible la profundidad testimonial, la singularidad cultural o el potencial transformador de muchas obras literarias en todo el mundo. Como señalaba Bassnett, «translation is not just the transfer of texts from one language into another, it is now rightly seen as a process of negotiation between texts and between cultures, a process during which all kinds of transactions take place mediated by the figure of the translator».[69] En la era de la automatización, esa figura mediadora, con su bagaje humanístico, ético y creativo, sigue siendo irremplazable.

Bibliografía

Abdelhalim, Safaa M., Asma A. Alsahil y Zainab A. Alsuhaibani. 2025. «Artificial Intelligence Tools and Literary Translation: A Comparative Investigation of ChatGPT and Google Translate from Novice and Advanced EFL Student Translators' Perspectives». *Cogent Arts & Humanities* 12.1 ‹https://www.tandfonline.com/doi/full/10.1080/23311983.2025.2508031›

Aixelá, Javier Franco. 1996. «Culture-Specific Items in Translation». En *Translation, Power, Subversion.* Román Álvarez y M. Carmen-África Vidal, eds. Clevedon UK, Multilingual Matters. 52-78.

Al Rousan, Rafat, Raghad Jaradat y Mona Malkawi. 2025. «ChatGPT Translation vs. Human Translation: An Examination of a Literary Text». *Cogent Social Sciences* 11.1 ‹https://www.tandfonline.com/doi/full/10.1080/23311886.2025.2472916›

---

[67] Van Egdom 2024, 14.
[68] Sheng y Yankun 2023.
[69] Bassnett 2002, 6.

Aldawsari, Hamad Abdullah H. 2025. «Evaluating the Performance of Large Language Models on Arabic Lexical Ambiguities: A Comparative Study with Traditional Machine Translation Systems». *World Journal of English Language* 15.3: 354-54.

Baron, Naomi S. 2023. *Who Wrote This? How AI and the Lure of Efficiency Threaten Human Writing*. Redwood City, Stanford University Press.

Bassnett, Susan. 2002. *Translation Studies*. New York, Routledge.

Castaldo, Antonio, Sheila Castilho, Joss Moorkens y Johanna Monti. 2025. «Extending CREAMT: Leveraging Large Language Models for Literary Translation Post-Editing». *Proceedings of the 20th Machine Translation Summit (MT Summit 2025)*. Geneva, European Association for Machine Translation. 506-15.

Declercq, Christophe y Gys-Walt Van Egdom. 2023. «No More Buying Cats in a Bag? Literary Translation in the Age of Language Automation». *Revista Tradumàtica. Tecnologies de la Traducció* 21: 49-62.

Đorđević, Jasmina P. 2024. «Gender Bias and Women's Rights in the Workplace: The Potential Impact of English-German Translation Tools». En *Gendered Technology in Translation and Interpreting: Centering Rights in the Development of Language Technology*. Esther Monzó-Nebot y Vicenta Tasa-Fuster, eds. New York, Routledge. 202-24.

Gao, Ruiyao, Yumeng Lin, Nan Zhao y Zhenguang G. Cai. 2024. «Machine Translation of Chinese Classical Poetry: A Comparison among ChatGPT, Google Translate, and DeepL Translator». *Humanities and Social Sciences Communications* 11. ‹https://www.nature.com/articles/s41599-024-03363-0›

García González, Marta. 2024. «The Role of Human Translators in the Human-Machine Era: Assessing Gender Neutrality in Galician Machine and Human Translation». En *Gendered Technology in Translation and Interpreting: Centering Rights in the Development of Language Technology*. Esther Monzó-Nebot y Vicenta Tasa-Fuster, eds. New York, Routledge. 173-201.

Gaspari, Federico, Antonio Toral, Sudip Kumar Naskar, Declan Groves y Andy Way. 2014. «Perception vs. Reality:

Measuring Machine Translation Post-Editing Productivity». *Proceedings of the 11th Conference of the Association for Machine Translation in the Americas*. Vancouver, Association for Machine Translation in the Americas. 60-72. ‹https://aclanthology.org/2014.amta-wptp.5›

Ghosh, Sourojit y Srishti Chatterjee. 2024. «Misgendering and Assuming Gender in Machine Translation When Working with Low-Resource Languages». En *Gendered Technology in Translation and Interpreting: Centering Rights in the Development of Language Technology*. Esther Monzó-Nebot y Vicenta Tasa-Fuster, eds. New York, Routledge. 274-90.

Gough, Joanna. 2024. «Individual Variations in Information Behavior of Professional Translators: Towards a Classification of Translation-Oriented Research Styles». *Translation Studies* 17.2: 394-415.

Guerberof Arenas, Ana. 2014. «Correlations between Productivity and Quality when Post-editing in a Professional Context». *Machine Translation* 28.3-4: 165-86.

Guerberof Arenas, Ana y Antonio Toral. 2022. «Creativity in Translation: Machine Translation as a Constraint for Literary Texts». *Translation Spaces* 11.2: 184-212.

Guerberof-Arenas, Ana y Antonio Toral. 2020. «The Impact of Post-Editing and Machine Translation on Creativity and Reading Experience». *Translation Spaces* 9.2: 255-82.

Hu, Kaibao y Xiaoqian Li. 2023. «The Creativity and Limitations of AI Neural Machine Translation: A Corpus-Based Study of DeepL's English-to-Chinese Translation of Shakespeare's Plays». *Babel* 69.4: 546-63.

Katan, David. 2023. «Tools for Transforming Translators into Homo Narrans or "What Machines Can't Do"». En *The Human Translator in the 2020s*. Gary Massey, Elsa Huertas-Barros y David Katan, eds. New York, Routledge. 74-90.

Kenny, Dorothy y Marion Winters. 2020. «Machine Translation, Ethics and the Literary Translator's Voice». *Translation Spaces* 9.1: 123-49.

Kenny, Dorothy. 2019. «Machine Translation». En *Routledge Encyclopedia of Translation Studies*. Mona Baker y Gabriela Saldanha, eds. New York, Routledge. 305-10.

Landers, Clifford E. 2001. «The Uniqueness of Literary Translation». *Literary Translation: A Practical Guide*. Bristol, Blue Ridge Summit, Multilingual Matters. 7-12.

Leonardi, Letizia. 2025. «Poetic Machine Translation: An Oxymoron? The Case of Giacomo Leopardi into English». En *Traducción y sostenibilidad cultural II: retos y nuevos escenarios*. Fernández, Sara García, Fátima Gómez-Cáneba, Beatriz Guerrero García, Francesca Placidi, Margarita Savchenkova y Susana Schoer-Granado, eds. Salamanca, Ediciones Universidad de Salamanca. 641-49.

Li, Bo. 2023. «Ethical Issues for Literary Translation in the Era of Artificial Intelligence». *Babel: Revue internationale de la traduction / International Journal of Translation* 69.4: 529-45.

Li, Xiaoye, Xiangling Wang y Wentian Lai. 2025. «The Usability of Neural Machine Translation in Creative-Text Post-Editing: Evidence from Users' Performance and Perception». *International Journal of Human-Computer Interaction* 41.3: 1-12.

Manapbayeva, Zhannura, Gulkhadisha Zaurbekova, Karlygash Ayazbekova, Aigul Kazezova y Kunsulu Pirmanova. 2024. «AI in Literary Translation: ChatGPT-4 vs. Professional Human Translation of Abai's Poem "Spring"». *Procedia Computer Science* 251: 526-31.

Massey, Gary, Elsa Huertas-Barros y David Katan, eds. 2023. *The Human Translator in the 2020s*. New York, Routledge.

Mirghaderi, Leilasadat, Monika Sziron y Elisabeth Hildt. 2023. «Ethics and Transparency Issues in Digital Platforms: An Overview». *AI* 4.4: 831-43.

Moorkens, Joss. 2023. «"I Am Not a Number": On Quantification and Algorithmic Norms in Translation». *Perspectives: Studies in Translation Theory and Practice* 32.3: 477-92.

Moorkens, Joss, Andy Way y Séamus Lankford. 2024. *Automating Translation*. New York, Routledge.

Moorkens, Joss, Antonio Toral, Sheila Castilho y Andy Way. 2018. «Translators' Perceptions of Literary Post-editing Using Statistical and Neural Machine Translation». *Translation Spaces* 7.2: 240-62.

Naeem, Arfa, Ayesha Saif ur Rehman y Arslan Rasheed. 2025. «Evaluating Cultural Adaptation in AI Translations: A

Framework and Implications for Literary Works». En *AI Applications for English Language Learning*. Muhammad Mooneeb Ali, Muhammad Nadeem Anwar, Shawana Fazal y Shazia Ayyaz, eds. Hershey, IGI Global. 223-52.

Nagy, Ladislav. 2025. «Challenges to Literary Translation in the Age of Technology». En *Traducción y sostenibilidad cultural II: retos y nuevos escenarios*. Fernández, Sara García, Fátima Gómez-Cáneba, Beatriz Guerrero García, Francesca Placidi, Margarita Savchenkova y Susana Schoer-Granado, eds. Ediciones Salamanca, Universidad de Salamanca. 651-57.

Opaluwah, Adeyola. 2025. «A Case for Machine in The Translation of Culture-Specific Items». En *Traducción y sostenibilidad cultural II: retos y nuevos escenarios*. Sara García Fernández, Fátima Gómez-Cáneba, Beatriz Guerrero García, Francesca Placidi, Margarita Savchenkova y Susana Schoer-Granado, eds. Ediciones Salamanca, Universidad de Salamanca. 659-71.

Ouldelhaj, Driss y Hajar Benmakhlouf. 2015. «La inteligencia artificial y la traducción literaria: realidad y perspectivas». *Perspectivas de la Comunicación* 18. ‹https://www.perspectivasdelacomunicacion.cl/index.php/perspectivas/article/view/3603›

Pan, Qiuyang, Qingyi Song y Tao Feng. 2024. «Comparative Study of English and Chinese Translation Theory: Revealing the Potential of Digital Technologies». *Translation Review* 118.1: 28-43.

Portelli, Sergio. 2024. «Translating the Unknown: A Case Study on the Usefulness of Machine Translation in Comparative Literature Research». En *Inclusion, Diversity and Innovation in Translation Education*. Alejandro Bolaños García-Escribano y Mazal Oaknín, eds. London, UCL Press. 19-35.

Rico Pérez, Celia y Antonio Jesús Martínez Pleguezuelos. 2024. «Exploring Gender Bias in Machine Translation of Legal Texts». En *Gendered Technology in Translation and Interpreting: Centering Rights in the Development of Language Technology*. Esther Monzó-Nebot y Vicenta Tasa-Fuster, eds. New York, Routledge. 253-73.

Rico Pérez, Celia. 2025. «Estudio del efecto de imprimación de la traducción automática en los textos institucionales de la Unión Europea traducidos al español». *Revista de Humanidades Digitales* 10: 48-72.

Robinson, Douglas. 2023. «Walter Benjamin as Translator as John Henry. Competing with the Machine». *Babel: Revue internationale de la traduction / International Journal of Translation / Revista Internacional de Traducción* 69.4: 499-528.

Sánchez-Torrón, Marina y Philipp Koehn. 2016. «Machine Translation Quality and Post-Editor Productivity». *Proceedings of the 11th Conference of the Association for Machine Translation in the Americas: MT Researchers' Track*. Association for Machine Translation in the Americas (AMTA): 16-26. ‹https://aclanthology.org/2016.amta-researchers.2›

Sharar, Hana, Aya W. Al-Zagha y Eman Ali. 2025. «ChatGPT in Translating Cultural Nuances in *O Night* by Gibran Khalil Gibran». *International Journal of Linguistics, Literature and Translation* 6.4: 30-38.

Sheng, Anfeng y Kong Yankun. 2023. «Neural Machine Translation and Human Translation: A Political and Ideological Perspective». *Babel* 69.4: 483-98.

Spoturno, María Laura. 2024. «Traducción literaria e inteligencia artificial: consideraciones para la formación universitaria». *Cadernos de Tradução* 44.1. ‹https://periodicos.ufsc.br/index.php/traducao/article/view/100602›

Stevenson, Angus y Christine A. Lindberg, eds. 2011. «Artificial Intelligence». *New Oxford American Dictionary*. Oxford University Press. ‹https://www.oxfordreference.com/view/10.1093/acref/9780195392883.001.0001/m_en_us1223122›

Toral, Antonio y Andy Way. 2014. «Is Machine Translation Ready for Literature?» *Proceedings of Translating and the Computer* 36: 174-76.

Toral, Antonio y Andy Way. 2015. «Translating Literary Text between Related Languages Using SMT». En *Proceedings of the Fourth Workshop on Computational Linguistics for Literature*. Anna Feldman, Anna Kazantseva, Stan Szpakowicz y Corina Koolen, eds. Denver, Association for Computational Linguistics. 123-32.

Toral, Antonio, Martijn Wieling y Andy Way. 2018. «Post-Editing Effort of a Novel with Statistical and Neural Machine Translation». *Frontiers in Digital Humanities* 5.

‹https://www.frontiersin.org/journals/digital-humanities/articles/10.3389/fdigh.2018.00009/full›

Van Egdom, Gys-Walt M. W. 2024. «Bridging Linguistic Divides? A Critical Exploration of Machine Translation's Role in Fostering Cross-Cultural Accessibility in Literature». *Íkala, Revista de Lenguaje y Cultura* 29.3. ‹https://revistas.udea.edu.co/index.php/ikala/article/view/356102›

Vanmassenhove, Eva. 2024. «Gender Bias in Machine Translation and the Era of Large Language Models». En *Gendered Technology in Translation and Interpreting: Centering Rights in the Development of Language Technology*. Esther Monzó-Nebot y Vicenta Tasa-Fuster, eds. New York, Routledge. 225-52.

Wang, Hongtao. 2023. «Defending the Last Bastion: A Sociological Approach to the Challenged Literary Translation». *Babel: Revue internationale de la traduction / International Journal of Translation / Revista Internacional de Traducción* 69.4: 465-82.

Wang, Lan. 2023. «The Impacts and Challenges of Artificial Intelligence Translation Tool on Translation Professionals». *SHS Web of Conferences* 163. ‹https://www.shs-conferences.org/articles/shsconf/abs/2023/12/shsconf_icssed2023_02021/shsconf_icssed2023_02021.html›

Wang, Shanshan, Derek Wong, Jingming Yao y Lidia Chao. 2024. «What is the Best Way for ChatGPT to Translate Poetry?». *Proceedings of the 62nd Annual Meeting of the Association for Computational Linguistics*. Bangkok, Association for Computational Linguistics. 14025-43.

Way, Andy, Andrew Rothwell y Roy Youdale. 2023. «Why More Literary Translators Should Embrace Translation Technology». *Tradumàtica. Tecnologies de la Traducció* 21: 87-102.

Winters, Marion y Dorothy Kenny. 2024. «Mark My Keywords: A Translator-Specific Exploration of Style in Literary Machine Translation». En *Computer-Assisted Literary Translation*. Andrew Rothwell, Andy Way y Roy Youdale, eds. New York, Routledge. 69-87.

Zhang, Ran, Wei Zhao y Steffen Eger. 2025. «How Good Are LLMs for Literary Translation, Really? Literary Translation Evaluation with Humans and LLMs». En *Proceed-*

ings of the 2025 Conference of the Nations of the Americas Chapter of the Association for Computational Linguistics: Human Language Technologies. Luis Chiruzzo, Alan Ritter y Lu Wang, eds. Albuquerque, Association for Computational Linguistics. 10961-88.

O direito autoral e os desafios da regulação da inteligência artificial no Brasil

Tiago Andreotti e Silva, José Paulo Gutierrez
e Ana Paula Martins Amaral

Introdução

A criatividade é uma característica inerente ao ser humano. Por meio dela, construímos obras dos mais variados tipos que passam a ser admiradas e consumidas por várias pessoas. Em algumas situações, a obra em si é de difícil reprodução, em razão dos passos para a sua materialização. Em outras situações, há uma facilidade imensa para a reprodução de um trabalho intelectual, o que acaba aproximando-o do conceito de bem público,[1] em que sua utilização pode ser reproduzida sem excluir outros e não há uma restrição ao seu uso. Porém, essa característica de «bem público» tende a afastar quem deseja se dedicar profissionalmente a criações que, por sua natureza, possuem essa facilidade em serem copiadas. De que adianta gastar meses, ou até anos, na criação de um livro, se o livro poderá ser facilmente copiado e vendido por terceiros, que não o autor do livro, que se beneficiarão de sua obra?

O racional para a criação de direitos autorais, sob o ponto de vista econômico, serve exatamente para lidar com essa questão, tendo por objetivo dar proteção à criação de uma obra e possibilitar que o seu autor possa explorar o seu trabalho economicamente, excluindo, em razão de uma norma jurídica, terceiros não autorizados para exploração de seu trabalho. Com o surgimento da inteligência artificial (IA) passamos a ter a criação de obras com menor intervenção humana. É verdade que a criação de determinado sistema de IA foi realizada por pessoas, porém, as obras resultantes de um comando dado a essa mesma IA acabam recebendo menor intervenção.

---

[1] Bens públicos são definidos como bens que o uso por uma pessoa não diminui o uso por outra (non-rivalrous) e não há a possibilidade de prevenção de seu uso (non-excludable), Reiss 2021.

Por exemplo, em menos de dois minutos eu pedi ao Gemini,[2] IA da Google, que elaborasse uma pintura, no estilo do quadro *O Grito*,[3] retratando uma capivara no Parque das Nações Indígenas, em Campo Grande/MS, que 'pintou' para mim a seguinte imagem:

Figura 1. A Capivara que grita
Fonte: Tiago Andreotti e Silva

Nominei a imagem de *A Capivara que grita*. Há, nesse caso, direito autoral? Para quem? A concepção da pintura foi minha, pois fui eu que pensei em retratar a capivara no Parque das Nações Indígenas, com inspiração no quadro *O Grito*, porém, foi uma máquina que realizou a pintura. Para tentarmos en-

---

[2] O comando dado foi o seguinte: «Elabore uma pintura, no estilo de 'o grito', retratando uma capivara no parque das nações indígenas, em Campo Grande/MS».

[3] *O Grito* (em norueguês: *Skrik*) é uma série de quatro pinturas do norueguês Edvard Munch, 1893. A obra representa uma figura andrógina num momento de profunda angústia e desespero. O plano de fundo é a doca do fiorde de Oslo (em Oslo) ao pôr do sol. [...] A série tem quatro pinturas conhecidas: dois dos quadros da série, *A Ansiedade* (em norueguês: *Angst*) e *O Desespero* (em norueguês: *Fortvielse*) encontram-se na posse do Museu Munch, em Oslo, outra na Galeria Nacional de Oslo e outra em coleção particular», Prelinger 2001.

tender as dificuldades em responder a essas questões, é necessário, primeiramente, entendermos o que é o Direito Autoral e quais são os limites para a proteção de obras, para, então, analisarmos as dificuldades inerentes à sua aplicação em criações intermediadas por IA. Assim, o capítulo está dividido em três partes: na primeira traremos breves considerações sobre o que é o direito autoral; na segunda verificaremos os limites de proteção de obras no contexto jurídico brasileiro para depois discutirmos as questões jurídicas relacionadas à proteção de obras criadas por IA.

O que são Direitos Autorais?

Os direitos autorais nada mais são que um conjunto de regras criadas para a proteção de criações literárias e artísticas. No Brasil, os direitos autorais são regulados pela Lei 9.610/98, que, em seu art. 7º, define que as obras intelectuais protegidas são «as criações do espírito, expressas por qualquer meio ou fixadas em qualquer suporte, tangível ou intangível, conhecido ou que se invente no futuro».[4] Nos EUA, o Copyright Act of 1976 define que a proteção do direito autoral é dada para «obras de autoria original fixadas em qualquer meio de expressão, conhecidos atualmente ou em momento posterior, que podem ser percebidos, reproduzidos, ou de outra forma comunicadas, diretamente ou com o auxílio de uma máquina ou equipamento».[5]

Percebe-se que, em ambos os ordenamentos jurídicos, há a necessidade de fixação da obra em um suporte, ou seja, não se protege somente ideias abstratas, mas sim ideias que conseguem ser expressas em determinado meio. Portanto, não se protege a «criação» pura e simples, mas sim a «obra», ou seja, a criação fixada em um determinado meio de expressão.[6] Os

---

[4] Lei 9.610/98, art. 7º.

[5] Section 102. Tradução livre. No original: «[c]opyright protection subsists, in accordance with this title, in original works of authorship fixed in any tangible medium of expression, now known or later developed, from which they can be perceived, reproduced, or otherwise communicated, either directly or with the aid of a machine or device».

[6] «A "obra intelectual" de que tratam as leis de Direitos Autorais configura uma criação humana concretizada em determinada forma, exteriorizada de alguma maneira e resultante do aporte individual ou da contribuição coletiva de determinadas pessoas. Outras criações humanas existem que não estão compreendidas na noção legal de obra, seja

direitos do autor têm um caráter dúplice – moral e patrimonial. Sob o aspecto moral, o autor poderá: reivindicar a autoria da obra; ter seu nome indicado ou anunciado como sendo o do autor, quando sua obra for utilizada; conservar a obra inédita; assegurar a integridade da obra; modificar a obra; retirar a obra de circulação ou suspender qualquer forma de utilização já autorizada, quando a circulação ou utilização implicarem afronta à reputação e à imagem; e ter acesso a exemplar único e raro da obra, para, por meio de processo fotográfico ou audiovisual, preservar sua memória.[7]

O aspecto patrimonial consiste em um direito exclusivo do autor em «utilizar, fruir e dispor da obra literária, artística ou científica»,[8] de forma que há a necessidade de autorização prévia e expressa do autor para a utilização de sua obra.[9] Ao permitir a exclusão de terceiros do uso de uma obra sem autorização, o que a norma de direito autoral faz é transformar um bem que, sob o ponto de vista econômico, era público, em um bem privado, criando incentivos econômicos para quem deseja criar obras, já que terá a exclusividade de uso do produto de sua criação.[10]

Por outro lado, alguns críticos do sistema de direito autoral afirmam que tais regras limitam o uso de obras que poderiam ser utilizadas por várias pessoas ao mesmo tempo, sem custo adicional, o que, economicamente, diminuiria o bem-estar da população em geral, já que aumentaria o custo de acesso à cultura. Também é importante mencionar outras críticas ao sistema de proteção ao direito autoral, que questionam se a proteção, da forma como é estruturada, de fato cria incentivos para os autores produzirem suas obras.[11] Porém, é importante ressaltar que cada país possui o seu sistema de proteção sobre o direito autoral, que pode ou não estar inserido no sistema internacional de proteção, estabelecido por meio de acordos multilaterais, como, por exemplo, o Acordo TRIPS, no âmbito

---

em virtude da tradicional dicotomia forma-conteúdo ou ideia-expressão, seja em face da natureza da criação que constitui objeto desta disciplina, distinguindo-a de outros ramos da Propriedade Intelectual», Santos 2020.

[7] Lei 9.610/98, art. 24, incisos I a VII.

[8] Lei 9.610/98, art. 28.

[9] Lei 9.610/98, art. 29.

[10] Mankiw 2021, 237.

[11] Sterk 1996.

da Organização Mundial do Comércio, e os tratados realizados sob os auspícios da Organização Mundial da Propriedade Intelectual.

Direitos Autorais no Brasil

O Brasil possui um sistema jurídico cuja origem histórica se encontra na tradição da «Civil Law», com raízes que podem ser traçadas à Revolução Francesa, onde a lei possuía maior importância do que as decisões judiciais para o fim de criação de direitos e obrigações, estando os juízes limitados a simplesmente a aplicarem, sem a possibilidade de interpretações que poderiam distorcê-la, o que é distinto da origem da «Common Law», onde os juízes possuíam poderes mais amplos, podendo controlar atos do parlamento.[12] Porém, com o desenvolvimento dos sistemas legais pelo mundo, hoje essa distinção entre os sistemas são cada vez menores, possuindo os juízes funções semelhantes em ambos.[13]

No Brasil, o sistema jurídico tem como fundamento maior a Constituição, cuja legitimidade decorre de uma «força política capaz de estabelecer e manter o vigor normativo»[14] de seu texto, à qual as leis devem obedecer a seus contornos formais e materiais.[15]

Em relação ao tema «Direitos Autorais», é possível identificarmos várias regras e princípios constitucionais que devem ser observados na criação e aplicação da legislação infraconstitucional que regula o assunto. Podemos mencionar a necessidade de, em respeito aos direitos autorais de terceiros na manifestação de pensamento e realização de atividade intelectual, artística e científica, termos o cuidado para, ao criar uma obra, não violar a intimidade, vida privada ou honra de pessoas, bem como realizarmos a ponderação entre o direito de acesso aos bens culturais e a função social da propriedade[16]

---

[12] Marinoni 2009, 26-28.
[13] Marinoni 2009, 39-40.
[14] Mendes e Branco 2024, 67.
[15] Mendes e Branco 2024, 72.
[16] A Constituição Federal, em seu art. 215, dispõe que «O Estado garantirá a todos o pleno exercício dos direitos culturais e acesso às fontes da cultura nacional, e apoiará e incentivará a valorização e a difusão das manifestações culturais», e em seu art. 5º, XIII, que «a propriedade atenderá a sua função social». Tais normas devem ser lidas não como uma autorização para afastar direitos autorais, mas sim para se buscar

com o direito de propriedade em si, que deverá ser respeitado para permitir a existência de incentivos para a criação de novas obras.[17] É dentro desse arcabouço constitucional que a Lei 9.610/98 foi criada, a qual passamos a analisar, para delimitar os principais aspectos de proteção ao direito autoral.

Conforme acima mencionado, obras protegidas são «criações do espírito, expressas por qualquer meio ou fixadas em qualquer suporte, tangível ou intangível, conhecido ou que se invente no futuro».[18] Não se enquadram no domínio da proteção de direitos autorais os objetos elencados no art. 8º da Lei, como ideias, procedimentos normativos, esquemas para atos mentais como jogos ou negócios, formulários e instruções, textos de tratados, convenções e leis, entre outros.

O autor, a quem é dada a proteção sobre a obra, «é a pessoa física criadora de obra literária, artística ou científica»,[19] considerando-se «autor da obra intelectual, não havendo prova em contrário, aquele que, por uma das modalidades de identificação [...] tiver, em conformidade com o uso, indicada ou anunciada essa qualidade na sua utilização»,[20] e que passará a ter «direitos morais e patrimoniais sobre a obra que criou».[21] A proteção da obra consiste no direito exclusivo de utilizar, fruir e dispor da obra literária, artística ou científica,[22] impondo a necessidade de autorização prévia e expressa do autor para a utilização da obra em qualquer modalidade.[23] Os direitos patrimoniais são transferidos aos seus herdeiros, e perduram por 70 anos a partir de 1º de janeiro do ano subsequente ao de seu falecimento.[24] Para obras anônimas ou pseudônimas, o prazo de 70 anos é contado a partir do 1º ano seguinte ao de sua primeira publicação.[25] A Lei 9.610/98 também prevê, em seu art. 46, algumas situações que não configuram ofensa aos direitos autorais, permitindo alguma liberdade na

---

soluções que permitam o desenvolvimento da cultura nacional conjuntamente com a possibilidade de acesso à mesma.

[17] Netto 2023, 252.
[18] Lei 9.610/98, art. 7º.
[19] Lei 9.610/98, art. 11.
[20] Lei 9.610/98, art. 13.
[21] Lei 9.610/98, art. 22.
[22] Lei 9.610/98, art. 28.
[23] Lei 9.610/98, art. 29.
[24] Lei 9.610/98, art. 41.
[25] Lei 9.610/98, art. 43.

utilização de obras protegidas, sem a necessidade de autorização por parte do Autor. São situações pontuais, que dificilmente comprometem a possibilidade de exploração comercial, pelo Autor, de sua obra.

Como consequência da violação de direitos autorais, a lei permite a apreensão de exemplares fraudulentamente reproduzidos (art. 102) e a suspensão ou interrupção de exibição de obras criadas com violação a direitos autorais (art. 105), a destruição de exemplares ilícitos e instrumentos utilizados para a sua criação (art. 106), além de impor o direito de indenização a quem tiver o seu direito autoral violado (art. 102). Sob o ponto de vista jurídico, as principais controvérsias estão relacionadas ao que é uma obra, para fins de proteção, quem pode ser autor, e se uma obra, criada com o auxílio de IA, pode ser atribuída a uma pessoa física, e, em caso positivo, se essa pessoa seria quem deu o comando para a máquina ou quem criou a IA.[26] Nesse contexto, é importante nos aprofundarmos um pouco mais no conceito de obra do sistema jurídico brasileiro, para fins de proteção de direitos autorais, bem como sobre os contornos para delimitar a autoria de uma obra.

A autoria, apesar das diferentes possíveis acepções, perpassa pela «análise da relação entre determinada criação intelectual e seu criador»,[27] sendo que sua função é tanto para se imputar uma relação de causalidade entre criação e originador, quanto para que seja possível ao originador se apropriar da criação.[28] No Brasil, conforme acima visto, somente a pessoa física pode ser autora. Em outras palavras, não se atribui a autoria para uma pessoa jurídica, havendo sempre a necessidade de se identificar a pessoa humana que realizou o ato de criação da obra e de sua fixação. É possível que as obras protegidas sejam criadas por mais de uma pessoa, e há a previsão de proteção legal atribuída a todos os coautores que dela participaram. Porém, a lei não considera como coautoria o mero auxílio, como a revisão, atualização, fiscalização ou direção de edição ou apresentação (art. 15, §2º). Por sua vez, a obra, enquanto criação intelectual materializada, será protegida, segundo a doutrina, quando estiver no âmbito literário, artístico

---

[26] Espirito Santo 2022, 1832-48.
[27] Santos 2020, 18.
[28] Santos 2020, 20.

ou científico, tiver originalidade e estiver dentro do prazo de proteção legal.[29]

A originalidade, para fins de proteção, refere-se não ao ineditismo da obra, mas sim à forma pela qual a ideia é apresentada, permitindo uma classificação entre obras absolutamente originais, que não decorrem de derivação de outras obras, e relativamente originais, que decorrem de outras obras.[30] Assim, adaptações, traduções, e outros tipos de obras derivadas também são protegidas, sem prejuízo dos direitos originais do autor da obra original. Por exemplo, para a realização de um filme sobre um determinado livro, o titular do direito autoral do livro deverá autorizar a referida adaptação, que, ao ser realizada, passará também a ter proteção legal. É nesse contexto que passamos à análise de possíveis questões referentes à proteção de obras criadas com o auxílio de IA.

A IA e as dificuldades para a proteção de obras criadas com o seu auxílio

São várias as definições sobre o que pode ser considerado IA, havendo uma relação direta com o funcionamento da inteligência humana. Por exemplo, no teste de Turing, IA é definida como a possibilidade de a máquina se comunicar com humanos sem revelar que não é humana, enquanto para Marvin Minky, IA é permitir que máquinas realizem tarefas que demandam inteligência humana.[31] Em linhas gerais, os mecanismos usados em aprendizagem de máquina são os supervisionados e os não supervisionados. No primeiro, o modelo é treinado com o oferecimento de dados de entrada e resultados esperados, com o objetivo de identificar padrões, enquanto no segundo a metodologia utilizada é de aproximação dos dados utilizados, em razão da inexistência de rótulos para os dados fornecidos.[32]

O Google Gemini, utilizado para criar a imagem no início deste capítulo, é um exemplo de modelo de linguagem multimodal, que pode trabalhar com textos, imagens, áudio e vídeo, entre outros tipos de dados, e está habilitado para lidar com

---

[29] Netto 2023, 125.
[30] Netto 2023, 126-27.
[31] Jiang 2022, 1-2.
[32] Jiang 2022, 5-8.

vários tipos de tarefas diferentes,[33] treinado com documentos da internet, livros, códigos, imagens, áudios e vídeos.[34] Assim, considerando que os sistemas de IA necessitam utilizar dados existentes para serem desenvolvidos, há uma primeira questão legal que surge referente à violação de direitos autorais de terceiros. No contexto da lei brasileira, a inclusão em base de dados de uma obra protegida depende de autorização (art. 29, IX), de forma que a utilização de obras para o treinamento de sistemas de IA depende de autorização, quando tais obras estão protegidas. Uma segunda questão legal, relacionada à primeira, é que as obras elaboradas por meio da intervenção de um sistema de IA utilizam informações, que podem ou não estar protegidas por direito autoral, e criam um produto novo, que pode ser bastante similar às obras protegidas por direito autoral, o que também geraria uma violação ao direito do autor.

A Google, vislumbrando a possibilidade de usuários de seus serviços serem demandados em razão de violações de direitos, criou uma obrigação contratual de indenizar os seus clientes caso haja a alegação, por terceiros, de violação de direitos autorais no uso de seus sistemas de IA generativa.[35] Isso,

---

[33] Carraro 2024.

[34] Anil and Borgeaud 2025, 5.

[35] «19. Generative AI Services. [...] i. Additional Google Indemnification Obligations. (i) Generated Output. Google's indemnification obligations under the Agreement also apply to allegations that an unmodified Generated Output from a Generative AI Indemnified Service using only Google Pre-Trained Model(s), a Fine-Tuned Google Model, or a Customer Adapter Model used with a Google Pre-Trained Model infringes a third party's Intellectual Property Rights. This subsection (i) (Generated Output) does not apply if the allegation relates to a Generated Output where: (1) Customer creates or uses such Generated Output that it knew or should have known was likely infringing, (2) Customer (or Google at Customer's instruction) disregards, disables, modifies, or circumvents source citations, filters, instructions, or other tools Google makes available to help Customer create or use Generated Output responsibly, (3) Customer uses such Generated Output after receiving notice of an infringement claim from the rightsholder or its authorized agent, (4) the allegation is based on a trademark-related right as a result of Customer's use of such Generated Output in trade or commerce, or (5) Customer does not have the necessary rights to the Customer Data used to customize or retrain the Fine-Tuned Google Model or Customer Adapter Model, or customize such Generated Output using a Generative AI Service. 'Generative AI Indemnified Service' means a Service or feature listed at ‹https://cloud.google.com/terms/generative-

por si só, pode representar um impedimento, no sistema brasileiro, a termos direitos sobre uma obra criada com o auxílio de IA. Se a obra decorrente do uso de IA for uma obra que viola direito autoral de terceiro, não haverá a sua proteção.

Porém, superando a questão do uso de obras protegidas para o treinamento de IA e a similaridade do resultado da obra criada com obras protegidas, há ainda uma terceira questão jurídica que é relevante para a nossa discussão. A simples inserção de um comando, decorrente de uma ideia do usuário, em um sistema de IA é suficiente para atribuir ao resultado a proteção de direito autoral? Seria um comando dado a uma máquina uma «criação de espírito», nos termos do art. 7º da Lei 9.610/98?

Não há, até o momento da entrega deste capítulo para publicação, uma resposta clara na legislação brasileira, e o tema tampouco foi abordado por decisões judiciais. Porém, há, na doutrina, o entendimento no sentido de que, para termos a proteção do direito autoral, é necessária a originalidade da obra, que significa que «a forma na qual as ideias são expressas deve ser uma criação original do seu autor»,[36] além da necessidade, acima discutida, de estar inserida no domínio literário, artístico ou científico.[37] Para o Superior Tribunal de Justiça, esse requisito da originalidade é a existência «de uma certa novidade, tanto da obra intelectual como do seu título».[38] Assim, sendo o autor pessoa física, e trazendo uma ideia em um formato novo, mesmo que auxiliado por IA, há a possibilidade de entendermos ser a pessoa que idealizou e inseriu os comandos para a criação da obra sua autora, o que atrairia a proteção legal.

Em uma discussão sobre pinturas elaboradas por IA em sua dissertação de mestrado, Andrezza Moraes, ao comentar

---

ai-indemnified-services›, where the use of such Service or feature is not provided to Customer free of charge», Google 2025.

(ii) Training Data. Google's indemnification obligations under the Agreement also apply to allegations that Google's use of training data to create any Google Pre-Trained Model utilized by a Generative AI Service infringes a third party's Intellectual Property Rights. This indemnity does not cover allegations related to a specific Generated Output, which may be covered by subsection (i) (Generated Output) above», Google 2025.

[36] Afonso 2009, 12.

[37] Afonso 2009, 13.

[38] Superior Tribunal de Justiça, Resp 1.311.629.

um sistema de IA cujo objetivo seria criar obras com o mínimo de intervenção humana, explica que lhe falta um elemento importante em seu processo artístico, pois o espaço criativo no qual está inserido é isolado, com pouco contexto social, ao contrário dos humanos, que são inspirados por povos, pessoas, lugares, conflitos e política.[39] Segundo a autora, há uma analogia a ser feita entre obras criadas por IA e fotografias, que, quando a tecnologia foi criada, não possuía proteção do direito autoral, e vieram a ser consideradas como arte somente em um segundo momento.

Nesse contexto, dois casos de direito comparado são interessantes para ilustrar essa discussão. No primeiro, o fotógrafo David Slater deixou uma câmera na natureza, com o autofoco ligado, e um dos macacos do grupo que ele estava documentando tirou uma «selfie».[40] A discussão seria se o fotógrafo teria proteção, em relação à 'selfie', do direito autoral. O Copyright Office, nos EUA, esclareceu suas regras e estabeleceu que fotos tiradas por animais não podem ter proteção legal, pois ausente o caráter de criação por uma pessoa humana.[41] Assim, se utilizarmos essa linha de raciocínio, há um argumento a ser feito no sentido de que a obra na qual há intervenção da IA, principalmente se a intervenção for substancial, não é produzida por uma pessoa humana, de forma que não haveria proteção legal a essas obras.

No segundo caso, que trata não diretamente de direito autoral, mas sim de patentes no âmbito do direito alemão, onde também há o requisito de autoria de pessoa para a concessão da proteção, a Corte Federal Alemã decidiu que, apesar de não ser possível ter uma IA como inventora, é possível a obtenção de registro de patente de invenção que teve a intervenção de

---

[39] Moraes 2022, 44.
[40] Balganesh 2017, 2-4.
[41] «As discussed in Section 306, the Copyright Act protects "original works of authorship". 17 U.S.C. § 102(a) (emphasis added). To qualify as a work of "authorship" a work must be created by a human being. See Burrow-Giles Lithographic Co., 111 U.S. at 58. Works that do not satisfy this requirement are not copyrightable. The U.S. Copyright Office will not register works produced by nature, animals, or plants. Likewise, the Office cannot register a work purportedly created by divine or supernatural beings, although the Office may register a work where the application or the deposit copy(ies) state that the work was inspired by a divine spirit», U.S. Copyright Office, § 313.2.

IA na sua criação.[42] Utilizando essa lógica, teríamos a possibilidade de proteção de uma obra criada com o auxílio da IA, garantindo à pessoa física que deu o comando para a IA o direito autoral sobre a obra realizada. Neste momento, a certeza que nós temos é a incerteza sobre o assunto. Há argumentos favoráveis e desfavoráveis, com base na legislação em vigor atualmente, para a proteção de obras criadas com o auxílio da IA.

Considerando a justificativa econômica para a proteção dos direitos autorais, bem como os critérios jurídicos da originalidade e da necessidade de autoria por pessoa humana, arriscaria dizer que quanto maior a participação humana na realização da obra, ou seja, quanto mais detalhado o comando dado para a IA e, quanto mais a obra tenha sido inicialmente elaborada pelo autor, maior a chance de termos o reconhecimento do direito autoral em obras que foram construídas com o auxílio da IA.

Considerações Finais

O direito do autor se desenvolveu sob o argumento de que seria necessário atribuir direitos aos autores para que eles tivessem incentivos para criar obras literárias, artísticas e científicas. No mundo atual, onde a informação é disseminada instantaneamente, essa proteção se torna ainda mais importante. Porém, com o desenvolvimento da IA, que está cada dia mais acessível, passamos a ter a possibilidade de criações de textos e imagens instantaneamente, com pouca intervenção do operador do sistema.

A questão é se esses textos e imagens são protegidos pela lei, pois há uma grande diferença na possibilidade de sua utilização. Por exemplo, se protegida, *A Capivara que grita* não poderia ser utilizada por mais ninguém, já que haveria a necessidade de minha autorização. Aliás, essa posição acarretaria outras consequências, pois eu poderia até mesmo alegar que uma pessoa que tenha dado o mesmo comando para a IA, criando uma imagem parecida, estaria violando o meu direito autoral. Por outro lado, se não há proteção, qualquer pessoa poderia utilizar essa imagem como bem entendesse, sem haver qualquer necessidade de autorização.

Porém, levando essa lógica ao extremo, e com essa facilidade de criação, eu poderia automatizar a criação de diversos

---

[42] Klos e Köllner 2024.

textos e imagens, que seriam criados com o auxílio da IA, por meio de um comando por mim dado, que acabaria me dando direitos autorais sobre inúmeras obras e permitindo que eu impedisse outras pessoas de criarem algo parecido, mesmo que eu não soubesse exatamente o que foi criado. Os limites do direito autoral em obras criadas com base em AI ainda serão bastante discutidos, tanto com base na legislação atual, em tribunais, como em propostas de modernização do Direito Autoral, para regular esses novos aspectos da criação intelectual que estão surgindo com a IA. Assim, com base em toda essa exposição, termino o capítulo perguntando a você leitor: teria eu direito sobre a imagem *A Capivara que grita*, ou estaria ela no domínio público?

Bibliografia

Afonso, Otávio. 2009. *Direito Autoral: Conceitos Essenciais*. Barueri, Manole.

Anil, Rohan e Sebastian Borgeaud. 2025. «Gemini: a Family of Highly Capable Multimodal Models». *Google*. ‹https://storage.googleapis.com/deepmind-media/gemini/gemini_1_report.pdf›

Balganesh, Shyamkrishna. 2017. «Causing Copyright». *Columbia Law Review* 117.1: 1-78.

Carraro, Fabrício. 2024. «O que é o Google Gemini e o que esse modelo de IA é capaz de fazer com exemplo prático». *Alura*. 23 maio 2024. ‹https://www.alura.com.br/artigos/google-gemini?srsltid=AfmBOoon8Jp1OEZWtiPRfQvRP-hRp08jwyTslmBsWm_RfOqxJDbDjKi6›

Espirito Santo, Alex, Thiago Domingos Marques, Breno Ricardo de Araújo Leite e Irineu Afonso Frey. 2022. «Direito autoral de criações feitas por inteligência artificial: diferentes percepções para o mesmo dilema». *Revista de Gestão e Secretariado* 13.3: 1832-48.

Google. 2025. «General Service Terms». *Google*. ‹https://cloud.google.com/terms/service-terms›

Jiang Yuchen, Xiang Li, Hao Luo, Shen Yin e Okyay Kaynak, 2022. «Quo Vadis Artificial Intelligence?». *Discover Artificial Intelligence* 2.4.

Klos, Mathieu e Malte Köllner. 2024. «Malte Köllner: "You Can File a Patent Application on an AI-assisted Invention"». *Juve-Patent*. 25 julho 2024. ‹https://www.juve-patent.

com/people-and-business/malte-kollner-you-can-file-patent-applications-on-ai-assisted-inventions-dabus›

Mankiw, Nicholas Gregory. 2021. *Princípios de microeconomia.* São Paulo, Cengage Learning.

Marinoni, Luis Guilherme. 2009. «Aproximação crítica entre as jurisdições de Civil Law e de Common Law e a necessidade de respeito aos precedentes no Brasil». *Revista da Faculdade de Direito* 49: 11-58.

Mendes, Gilmar Ferreira e Paulo Gonet Branco. 2024. *Curso de Direito Constitucional.* São Paulo, SaraivaJur.

Moraes, Andrezza. 2022. «Inteligência Artificial e Direito Autoral: pinturas produzidas por IA e legal framework para uma lege ferenda». Dissertação de Mestrado em Direito. Porto Alegre, Universidade do Vale do Rio dos Sinos.

Netto, José Carlos Costa. 2023. *Direito autoral no Brasil.* São Paulo, Saraiva Jur.

Prelinger, Elizabeth. 2001. *After the Scream: The Late Paintings of Edvard Munch.* New Haven, Yale UP.

Reiss, Julian. 2021. «Public Goods». *Stanford Encyclopedia of Philosophy.* ‹https://plato.stanford.edu/entries/public-goods/#DefiPublGoodDistBetwDiffKindPublGood›

Santos, Manoel J. Pereira, Wilson Pinheiro Jabur and José de Oliveira Ascensão. 2020. *Direito autoral.* São Paulo, Saraiva Educação.

Superior Tribunal de Justiça. 2017. «Resp 1.311.629».

Sterk, Stewart E. 1996. «Rhetoric and Reality in Copyright Law». *Michigan Law Review* 94.5:1197-1249.

U.S. Copyright Office. 2021. *Compendium of U.S. Copyright Office Practices.* 3rd ed.

Wikipedia. 2025. «O Grito». *Wikimedia.* ‹https://pt.wikipedia.org/wiki/O_Grito›

The Talmud's Lore, Forevermore: Issues of Plagiarism and Professionalism in the Use of ChatGPT for Creative Writing Projects

Robert Simon

Rooted in the need for off-site teaching models during the COVID-19 Pandemic, the landscape of Post Pandemic teaching has become known as one of technological innovation and a growing dependence on artificial intelligence (AI) for every aspect of teaching and learning:

> Use of synchronous learning and technology-as-utility framework to underpin remote learning during the pandemic was intended to retain most of physical contact learning traditions. Teachers and students met synchronously and simultaneously online for learning to occur. IHEs safeguarded their operational efficiency to minimize the disruptive nature of remote learning.[1]

In responding to the growing use of ChatGPT to supplement and nurture the writing process in our students, many times the app's seemingly vast knowledge of text, interpretation, and application of concepts lead students to wholesale and unadulterated use of the app for their academic and creative works. Recent research has outlined the harmful effects of misuse of the app[2] as well as its appropriate applications in various fields of study.[3]

As debates rage in almost every academic field, in areas of creative activities and / or the expression of personal judgments, this new technology strikes a particularly sour chord. To understand how, if at all, an algorithm truly understands the human condition, perhaps even replicating or replacing traditional notions of discovery and creativity with the superficial copy and paste functionality AI provides, we first need to push its limits in and out of the classroom.

---

[1] Pitso 2023, 40.
[2] Abbas 2024, 2.
[3] Al-Huwaiti and Al-Malloudh 2024, 1.

In April of 2024, ChatGPT wrote the following when I prompted it to «write a poem about love and the Talmud»:

In the pages of the Talmud, love's tale unfolds,
A narrative of hearts in scriptures old.
Lines etched with wisdom, ancient and wise,
Echoing truths that in love's realm lie.
In love's embrace, souls intertwine,
Like verses weaving, in rhythm divine.
Mysteries unravel, in each tender touch,
As hearts whisper secrets, they cherish so much.
Love, like Talmudic text, layered and deep,
In its complexities, its promises keep.
Guiding souls through life's winding road,
In joy and sorrow, its teachings bestowed.
For love, like Torah, is a guiding light,
In the darkest hours, it shines so bright.
Binding hearts in sacred bond,
Through trials and triumphs, it responds.
In the Talmud's verses, love finds its voice,
In every chapter, it makes its choice.
To cherish, to honor, to hold and protect,
In the journey of love, its lessons reflect.
So let us learn from Talmud's lore,
The essence of love, forevermore.
With patience and kindness, let our hearts align,
In the eternal dance of love's design.

I then prompted it to «escrever um poema sobre Lisboa e um encontro amoroso». This is what it offered:

Em Lisboa, onde o sol se inclina sobre o rio,
Onde os azulejos contam histórias de desvario,
Num encontro de amantes, num cenário a sorrir,
Do amor e da paixão, vamos deixar fluir.
Pelos becos estreitos do Bairro Alto a vagar,
Entre fados e poesia, vamos nos encontrar.
Sob o brilho das estrelas, no miradouro a sonhar,
Nossos corações se entrelaçam, a se apaixonar.
O Tejo, testemunha silenciosa do nosso encontro,
Reflete o brilho nos olhos, num desejo pronto.
Num café da Baixa, entre risos e olhares,
A magia de Lisboa nos envolve a amar.
Nas ruas de Alfama, ao som de guitarras a soar,
Dançamos juntos, como se o tempo parar.
Num abraço apertado, o mundo desvanece,
E só ficamos nós dois, num amor que enaltece.
Em cada esquina, Lisboa é cúmplice do nosso querer,
Neste labirinto de emoções, onde nos perdemos a crescer.
Num encontro amoroso, nesta cidade encantada,
Lisboa nos une, numa história apaixonada.

As the reader can see in the examples above, this overly formalized structure and stereotypical imagery gives the output of both requests a stale and vapid feeling. Any knowledgeable reader will see clearly the shameless lifting of verses from known fado songs; poetic translations of Rumi, Rashi, and Rambam; Fernando Pessoa, etc. If not for the problem of the growing student addiction to ChatGPT, I would find these examples laughable.

Yet there is little to laugh about when instructing a writing course whose poetry assignments are chalk (or, should I say, megabyte) full of such examples. In the Fall of 2023 and Spring of 2024, I instructed a World Literatures course in which students were required to visit a site, give a detailed description of that site, find a thematic point of association / comparison with a work of literature studied in the most recent module, and finally, opine on their own learning from literary and real-world experience. This seems like a difficult task for ChatGPT to unravel, and in some cases, it turned out even worse than assumed.

Although most students complete their work independently, a few each semester chose to leave the entire assignment to the app's posthuman discretion and supposed wisdom. What followed were, in essence, prosaic versions of the superficial and fatuous examples above. Descriptions were ripped from online introductions, literary analyses generally did not fit the work in question (as I chose lesser-known works of poetry and prose for all but two cases, ChatGPT could not find much information on them and thus, I assume, resorted to retrieving information from similarly titled, yet utterly unrelated, works), and opinions came born of self-referential gibberish. The nature of ChatGPT's AI learning matrix explains this erroneous, yet highly self-confident, mode of expression, as can be seen in a published study of Ukrainian students' unintended misuse of the technology:

> The issue of combating plagiarism has become a very acute one for Ukrainian education in the twenty-first century. A new challenge is a chatbot that generates plagiarism-free student papers. It paraphrases the infor-

mation it finds, thus avoiding easily detectable borrow-
ings. This will require a rethinking of the standard system of knowledge testing and assessment.[4]

One student was apparently so enamored with the response they received, they neglected to remove the tell-tale
phrase, «As an AI language model, I don't possess feelings or
personal opinions»[5] from their report. The irony of technology's inherent honesty when facing a dishonest use of that
technology should not escape us. However, it should concern
us in the long term.

Given the cases described above, it should be evident that
instructors cannot approach students with a laissez faire attitude when offering the opportunity to seek out and use (or
misuse) such seemingly facile technologies. As Stepanenko and
Stupak state, «[i]n general, while students can benefit from
ChatGPT, they need guidance on how to use it in order to fully
understand a particular topic, think logically, and evaluate the
performance of the GPT when performing certain tasks».[6]

Yet the questions of copyright infringement and plagiarism, which current scholarship does not necessarily treat outside of the academic sphere, also loom large. What will happen
when a professional writer utilizes the app and, in doing so,
unwittingly cites an easily searchable source? This contribution will look to explore these challenges to the creative endeavor insofar as the pedagogical lens may or may not encap-
sulate its usage.

In terms of deeper academic criticism on the subject, a
plethora of voices beginning in 2021 has made both the opportunities and detriments of this new technology evident. Fuchs
discusses some of the concrete issues facing instructors where
AI use, whether licit or not, is involved.

On the positive side we see the potential for a rapid, personalized form of feedback for students:

> NLP models can provide students with personalized
> learning experiences by generating content tailored
> specifically to their individual learning needs. For ex-
> ample, when a student submits a response to a question,
> the model can analyze the response and provide feed-
> back customized to the student's understanding of the

---

[4] Stepanenko and Stupak 2023, 204.
[5] Response generated by ChatGPT.
[6] Stepanenko and Stupak 2023, 207.

material. This feedback can help the student identify areas where they might need additional support or where they have demonstrated mastery of the material. Furthermore, the processing models can generate customized learning plans for individual students based on their performance and feedback. These plans may include additional practice activities, assessments, or reading materials designed to support the student's learning goals. By providing students with these customized learning plans, these models have the potential to help students develop self-directed learning skills and take ownership of their learning process.[7]

Despite these helpful aspects of AI-based learning, the consequences can also take center stage and overpower through its lack of accuracy and dependence-forming potential:

The accuracy of the system depends heavily on the quality, diversity, and complexity of the training data, as well as the quality of the input data provided by students. [...] If the training data is not adequately diverse or is of low quality, the system might learn incorrect or incomplete patterns, leading to inaccurate responses. [...] the potential risk of overreliance on technology, which could undermine the development of important critical thinking skills (while critical thinking has been singled out as an exemplary skill, the list of skills is countless and multiple higher-order thinking skills could be further discussed in the context of chatbots). While these models can offer valuable support and personalized learning experiences, students must be careful to not [sic] rely too heavily on the system at the expense of developing their own analytical and critical thinking skills.[8]

As an example, when asking ChatGPT «Why was Cervantes considered father of the modern novel?», the response received included several valid points, such as the complex narrative structure and Humanist perspective (which was not necessarily new given that the Picaresque Novel had been circulating as a genre in Spain since the 1550s), yet it also mentioned the following:

Narrative Innovation: Don Quixote introduced a more self-aware and playful narrative. The story itself blurs the lines between reality and fiction, as Don Quixote, the protagonist, often mistakes the world around him for

[7] Fuchs 2023, 2.
[8] Fuchs 2023, 2.

the chivalric world of knights and quests. This narrative approach, which challenges the reader's perception of reality, would influence modern writers like James Joyce and Franz Kafka.[9]

What is the issue? On the surface this statement seems reasonable. The problem begins when you realize that all these elements, the blurred perspectives on reality, the rewriting of visual stimuli as anachronism, had existed in Western literature, and in fact, since Biblical writings (keeping in mind the anachronistic memories feeding many of the prophecies of the Nevi'im, for example). Also, if Joyce and Kafka drew from Cervantes, the effect happened quite indirectly. Flaubert, Johnson, and Ibsen were Joyce's true influences, at least in his initial writings; Jewish literatures and many Eastern European writers exercised a strong influence on Kafka.

As we see from these comments, as well as the example provided, we can get a glimpse of some of the major benefits, and pitfalls, of AI-generated feedback and real-time use. The issue becomes even more acute in the area of World Language composition writing. Fangchen discusses this issue within the limits of existing studies in terms of how «learners' technological anxiety could be transferred through AI learning based on the learning readiness of AI [coming to the conclusion that] blindly using AI in education might not improve learners learning outcomes and mental abilities and might even have the opposite effect».[10] In other words, instructors who act under the assumption that AI is designed for their specific learners' needs will find their students could, in reality, lose real knowledge for dependence on AI and on inaccurate knowledge it provides. As such, we observe a need for slow and methodical introduction of AI into the classroom setting, rather than the all-at-once approach many learning environments have taken:

> With the advancement of education informatization, many schools have adopted AI tools to support EFL teaching, and students have already chosen AI tools for learning. Teachers and students needed necessary support and scientific guidance as AI was gradually integrated into teaching and learning. Otherwise [*sic*] they

---

[9] Response generated by ChatGPT.
[10] Fangchen 2024, 133.

might not be able to notice potential problems or understand the problems that have arisen, which might further influence their long-term use of AI tools.[11]

The unfortunate reality of the introduction of instructional technologies in the contemporary classroom tends not to correspond to the statement above. In fact, we may consider general knowledge that, due in large part to the confluence of the COVID-19 pandemic and the post-pandemic stresses on educational institutions with leaps in AI, the tools enumerated in this essay and elsewhere were thrusted up instructors and students with little or no preparation for their responsible usage and eventual outcomes.

As we have observed so far, the use of AI has gone relatively unchecked at the classroom level, despite the plethora of studies warning against its blatant and widespread misuse by students. This extends into all aspects of academic training and its application in post-collegiate scenarios. In relation to a bedrock case of fair use of Andy Warhol paintings out of context, a similar case based on use of AI highlights some key issues for present and future use of the more recent technology:

> The non-expressive fair use doctrine is based on two principles that validate extensive, unauthorized copying conducted by machines. The initial principle asserts that machine-based consumption of copyrighted expression is not inherently infringing. If mechanical ingestion does not enhance human interaction with expressive works, it qualifies as non-expressive. The second premise suggests that these uses do not significantly impact copyright owners' markets, as their rights do not cover non-expressive components engaged in computerized analysis and value derivation.
>
> However, the rise of machine learning challenges these assumptions. Firstly, machine learning enables computers to extract value from expressive aspects of works, not just factual elements. Consequently, these uses might cease to be non-expressive. Secondly, machine learning could introduce a novel market threat: instead of merely replacing individual works, expressive machine learning might entirely substitute human authors, disrupting market impact assessment.[12]

---

[11] Fangchen 2024, 147.
[12] Chandrakhar 2024, 55.

Recent developments in legal interpretations of AI fair use in wholesale copying of available and protected online content allow for certain exceptions: «Use for Sophisticated Public Welfare Purposes», «Output based on AI Training and Content Generation», and «Data Laundering, a phenomenon of transforming pilfered data for legitimate applications».[13] Although in only exceptional cases would any of these situations allow for fair use rules to protect AI-obtained information, we may assume, given criticism on the subject cited above, that most users would be unaware of either these court cases or the minutia of legally permitted use of materials. Considering the poems and other texts cited here from my own trial run of ChatGPT, we may deduce that AI clearly does not know either.

A study led by Katsantonis has offered a more positive interpretation of the post-baccalaureate AI use experience:

> a large percentage of agreement with the statement that AI will make daily lives more convenient, acknowledging the relevance of AI for both daily life and the future necessity of AI. About half the students also expressed an interest in AI and indicated that they expected to use AI in their future profession [...] At the same time, the findings raise awareness of the fact that educators need to increase students' interest and engagement with AI since only half the students were interested in AI.[14]

The notion that students enjoy the availability and flexibility of AI tools comes as no surprise. The concomitant idea that the division between perceived usefulness and practical use in the classroom could potentially foment the student/instructor divide so common in the classroom.

Other studies reveal serious concerns about AI use in industries where human bias could infiltrate results:

> Despite its many benefits, there are concerns about the potential risks associated with Chat GPT. One of the main challenges is that bias could be introduced into the model, as the data used could contain biases or stereotypes. In addition, misuse of AI technology is a concern, such as the creation of fake content or the spread of misinformation.[15]

---

[13] Chandrakhar 2024, 57-59.
[14] Katsantonis 2024, 10.
[15] Graefen and Fazal 2024, 44.

These issues may also foment discord between instructors and AI tools. In a very essential way, faculty perceptions also play an integral role in the interpretation of AI use in student learning. The case of faculty perspectives in Pakistani universities offers a glimpse into the complicated optics associated with AI in the university environment:

> Teachers expressed both worries and positive points of view, resulting in a balanced approach. The concerns included maintaining academic rigor, avoiding excessive dependence on Chat GPT, and distinguishing between student and AI-generated material. These results are consistent with previous studies.[16]

This consistency seems to exist in contrast to another of the study's findings:

> The study found substantial negative association between instructor views and their perceptions of Chat GPT's influence upon students' learning experiences and academic achievement. This implies that the instructors with more optimistic attitudes were more likely to anticipate good results [...] results revealed a significant difference in novice and experienced instructors' perceptions of the drawbacks of the Chat GPT integration.[17]

This unique addition to the general findings of studies across countries and cultures indicates that the amount of experience, and overall disposition of instructors, with the technology have a clear influence on perceptions of its usefulness and appropriateness in the classroom setting. According to the same study, these function independently of each other.

Beyond the university, the ultimate measure of AI's place in the wider world is in the professional, post-collegiate context. Lopezosa and his team conducted interviews with journalists on the use of AI, and in particular, on the use of a newer generation of AI tools. Their findings pointed to a robust use of AI generated content and a recognized need for care taken precisely in the areas mentioned in other scholarship cited above:

> Journalism has been significantly impacted by AI for several years now, particularly in areas such as automated text generation for weather information, sports

---

[16] Kanwal 2023, 1.
[17] Kanwal 2023, 1.

results, and financial updates. AI is also utilized in relation to reader engagement and content recommendations, among other uses. The challenges presented by AI
and the importance of its use in a supervised and transparent manner is noted, emphasizing the idea of com-
plementarity rather than substitution. [...] Lastly, it is
noted that generative AIs can make errors, particularly
when used for data gathering, highlighting the need for
human involvement and robust verification processes
in AI-based journalism. In terms of the opportunities
presented by AI in journalism, they rely on media or-
ganizations' ability to harness AI effectively to
strengthen journalistic values. AI is already being employed for content verification and even to promote
more ethical communication products, such as detect-
ing self-reporting or gender bias. Notably, AI can be
used to identify discrepancies in sentence lengths attributed to men and women or to analyze the gender
representation in media by counting the use of images
featuring women and men.[18]

These aspects of AI use in the professional sphere also
come with warnings about appropriate instruction of AI use in
the journalism classroom,[19] most of which we have commented on previously. Given Katsantonis' findings, we must
acknowledge the difficulties of balancing faculty perception,
student perception, professional perception, and actual performance of AI.

AI as a tool for feedback production may have more productive potential in all field of study, thus helping to avoid the
pitfalls of overuse in student content production which, if unchecked, could have serious repercussions on post-collegiate
career development. Sun et al. explain one approach particularly apt for students in computer programming:

A concern that emerged during our investigation was
the possibility that students might employ ChatGPT or
comparable resources inappropriately and cite information from their work [...] [s]cholars have emphasized
the need for transparent communication and explicit
protocols regarding the implementation of AI assistance
to enhance the learning experience while upholding rigorous academic criteria.[20]

---

[18] Lopezosa 2023, 5.
[19] Lopezosa 2023, 7-9.
[20] Sun 2024, 16-17.

This conclusion mirrors the practices suggested previously here, adding that:

> In addition, during the ChatGPT-facilitated program-
> ming process, instructors need to integrate effective
> strategies for utilizing ChatGPT, and they should also
> monitor students closely for instances of academic dis-
> honesty, such as plagiarism when using ChatGPT or
> other web resources [...] [b]y improving the accuracy
> and consistency of the code generated, ChatGPT can be-
> come a more reliable and valuable tool for program-
> ming learning.[21]

The legal ramifications of AI also deserve attention in the sense that, to avoid the pitfalls of AI use indicated above, certain limitations of use have been proposed. The question of «whether compliance is feasible»[22] then takes center stage: «The optimal design of AI regulation is fundamentally different when technical and institutional constraints, both critical to compliance, are considered. Failure to do so will risk, [...] at worst, counterproductive or perverse downstream consequences?».[23] For example, the idea that somehow an agency can govern the use of AI in the professional and collegiate environments through regular monitoring sounds feasible:

> AI audits suffer from a number of technical feasibility
> constraints. First, there is a significant gap between the
> types of values and AI principles regulators envision au-
> dits measuring – e.g., privacy, robustness, or transpar-
> ency – and the existing methods for evaluating those
> values and principles in AI systems. Second, the sophis-
> tication of AI systems and their integration into complex
> software systems can make audit execution intracta-
> ble.[24]

From a real-world standpoint, the governing of AI is not realistic. Although Guha recommends a system of self-auditing on the part of AI producers with some government oversight,[25] the obvious difficulties of attempting to control such a widespread and notoriously imperfect technology evidence another layer of complexity in its appropriate use in any context.

---

[21] Sun 2024, 18.
[22] Guha 2024, 1478.
[23] Guha 2024, 1478.
[24] Guha 2024, 1536.
[25] Guha 2024, 1555-56.

In this, the use of AI for creative purposes based on academic history and usage, legal precedent, and overall dispositions toward AI use point in some unique directions. Returning to the examples provided at the beginning of this brief discussion, there exists a potential for expansion and refinement of human endeavors, especially in more technical fields where high levels of precise calculation meet more subjective needs. However, in more humanist fields, such as journalism, the use of AI as a tool has a more complex and wary reception, with mixed results in its use on behalf of original writing.

ChatGPT reached its own conclusions when I asked it the fundamental question posed here, «What are the potential effects of AI in the workplace»:

> AI has the potential to revolutionize the workplace, improving productivity, safety, and efficiency. However, its integration also presents challenges, including job displacement, ethical concerns, and the need for reskilling. Balancing the benefits with the risks will require careful planning and collaboration between businesses, workers, and policymakers.[26]

We can agree or disagree with the statements above. In researching this topic, I find myself both appalled at the technology's ability to synthesize its own thin-skinned illness in front of me: the lovely shell with the sickened and meaningless heart, as Ana Merino suggests in the title of the first section of her poetic collection *Curación* (*Cure*), *La piel de los enfermos* (*The skin of the ill*). «Reskilling» an «balancing» sound like attractive ideas, however the work done on the topic of collegiate and post-collegiate use of, and confidence of use in, AI reveal a more complex and multifaceted process of acceptance and regulation all the part of all involved. The kicking-of-the-can implied in the AI-generation response above by no means absolves, much less forgets, the human users of AI in their assumed or real acceptance of its factual or logical errors, and even less so the user's inability to perceive the ethical implications of such use. In the academic environment and in the workplace, writers engaging with AI will need to use a level of judicious, critical thinking to which we will have to become accustomed quickly. As post-Pandemic medical professionals at-

---

[26] Response generated by ChatGPT.

tempt to head off the long-term consequences of COVID-19 infection and prevent a future illness of this magnitude, we must also look to find healthy ways of incorporating AI in our academic and professional writing.

Bibliography

Abbas, Muhammad, Farooq Ahmed Jam and Tariq Iqbal Khan. 2024. *International Journal of Educational Technology in Higher Education* 21.10: 1-22.

Al-Huwaiti, Monah bint Saleh and Hessa bint Mohammad Al-Malloudh. 2024. «The Reality of Enhancing the Teaching of Social Studies through the Chat GPT Application from the Perspective of Social Studies Teachers in the Asir Region». *Journal of Educational Sciences and Human Studies* 37: 32-57.

Chandrakar, Vaishnavi. 2024. «From Canvas to Code: Analysing the Generative AI and Fair Use through the Lens of the Andy Warhol Verdict». *Journal of Intellectual Property Studies* 8.1: 47-60.

Fangchen, Wen, Yushun Li, Ying Zhou, Xin An and Quinghua Zou. 2024. «A Study on the Relationship between AI Anxiety and AI Behavioral Intention of Secondary School Students Learning English as a Foreign Language». *Journal of Educational Technology Development & Exchange* 17.1: 130-54.

Fuchs, Kevin. 2023. «Exploring the Opportunities and Challenges of NLP Models in Higher Education: is Chat GPT a Blessing or a Curse?». *Frontiers in Education* 8.1166682: 1-4.

Graefen, Bahar and Nadeem Fazal. 2024. «From Chat Bots to Virtual Tutors: An Overview of Chat GPT's Role in the Future of Education». *Archives of Pharmacy Practice* 15.2: 43-52.

Guha, Neel, Christie Lawrence, Lindsey A. Gailmand, Kit Rodolfa, Faiz Surani, Rishi Bommasani, Inioluwa Raji, Mariano-Florentino Cuéllar, Colleen Honigsberg, Percy Liang and Daniel E. Ho. 2024. «AI Regulation Has Its Own Alignment Problem: The Technical and Institutional Feasibility of Disclosure, Registration, Licensing, and Auditing». *George Washington Law Review* 92.6: 1473-557.

Kanwal, Ayesha, Syeda Khadija Hassan and Iffaf Iqbal. 2023. «An Investigation into How University-level Teachers Perceive Chat-GPT Impact upon Student Learning». *Gomal University Journal of Research* 39.3: 1.

Katsatonis, Argyrios and Ioannis G. Katsatonis. 2024. «University Students' Attitudes toward Artificial Intelligence: an Exploratory Study of the Cognitive, Emotional, and Behavioural Dimensions of AI Attitudes». *Educational Science* 14.988: 1-14.

Lopezosa, Carlos, Lluís Codina, Carles Pont-Sorribes and Mari Vállez. 2023. «Use of Generative Artificial Intelligence in the Training of Journalists: Challenges, Uses and Training Proposal». *Profesional de la información* 32.4: 1-12.

Merino, Ana. 2010. *Curación*. Madrid, Cátedra.

Pitso, Teboho. 2023. «Post-COVID-19 Higher Learning: Towards Telagogy, a Web-Based Learning Experience». *IAFOR Journal of Education* 11.2: 39-59.

Rincón Castillo, Alejandro Guadalupe, Giovanna Jackeline Serna Silva, Javier Pedro Flores Arocutipa and Haydeé Quispe Berrios. 2023. «Effect of Chat GPT on the Digitized Learning Process of University Students». *Journal of Namibian Studies* 33: 1-15.

Stepanenko, Olena and Olga Stupak. 2023. «The Use of GPT Chat [*sic*] among Students in Ukrainian Universities». *Scientific Journal of Polonia University* 58.3: 202-07.

Sun, Dan, Azzeddine Boudouaia, Chengcong Zhu and Yan Li. 2024. «Would ChatGPT-Facilitated Programming Mode Impact College Students' Programming Behaviors, Performances, and Perceptions? An Empirical Study». *International Journal of Educational Technology in Higher Education* 21.14: 1-22.

O ChatGPT é um fingidor: o ilusionismo dadaísta na produção 'literária' generativa

Marcelo Pacheco Soares

Em relativamente recente artigo no jornal *Folha de São Paulo*, intitulado «Quem são os intelectuais bilionários que preparam a ruptura apocalíptica de Trump», de 19 de abril de 2025, o filósofo Martim Vasques da Cunha, ao analisar as relações do atual governo federal estadunidense com alguns nomes oriundos do Vale do Silício e suas pretensões de poder, aponta que «a IA é a nova bomba atômica, um poder que contém a violência inevitável do Anticristo — simbolizado, nessa perspectiva, pela ordem democrática liberal dos últimos 70 anos».[1] Sem adentrarmos as teorias conspiratórias que movem os atores dessa jornada descrita por Cunha, surpreende aqui a consciência confessada dos detentores do poder de que a chamada Inteligência Artificial é capaz de moldar o pensamento alheio, a partir da ilusão de que poderia elaborar um raciocínio próprio e independente. Ora, se o mundo ocidental foi por décadas e pouco a pouco educado e construído sob fundamentos pautados em valores democráticos e liberais, a IA parece ser, para seus entusiastas nesses específicos contexto e uso, o meio mais eficaz de rapidamente minar seus alicerces e provocar uma (já em curso, ao que parece) ruptura ideológica severa, uma virada intelectual e comportamental que faça nascer na sociedade pensamentos outros, os quais questionem esse *status quo* e torne palatável novamente ideias que pareciam sepultadas desde as tragédias provocadas por aplicações de conceitos tais como, por exemplo, os que consolidaram regimes fascistas ou nazistas há cerca de um século.

Sem querer fazer *tabula rasa* de um processo complexo, podemos dizer que um dos muitos fatores que tornam possível tal fenômeno seria o de uma espécie de espelhamento, ou seja, a concepção de que há, do outro lado de uma tela ou coisa que o valha (através do espelho, como num mundo aliciano?), um ser que produza ou guarde ideias semelhantes às de seu inter-

---

[1] Cunha 2025.

locutor e vá, a partir dessas relações de identidades, manipulando-o para fazê-lo alcançar uma determinada escolha: seja para comprar um produto, votar em um político, introjetar um dogma religioso ou — como na alegoria do filme de Christopher Nolan de 2010 — «to *Inception* an idea» (no caso discutido no artigo de Cunha, uma ideia de caráter ideológico). E, insistamos na palavra, isso ocorre através de uma «ilusão», a de que a IA produziria um pensamento autônomo, quando, na verdade, o mesmo pensamento já programado previamente em si por uma mente humana — o engenheiro — é produzido na vítima também humana, mas a máquina mesma, mera intermediadora ou multiplicadora desse pensar, não é capaz de criar por si raciocínio algum de fato. A IA, afinal de contas, não raciocina, a Inteligência Artificial não é dotada da capacidade de *intelligentia*, do latim, *inter* > entre + *legere* > entender ou ler, daí sua artificialidade.

Trazemos esse preâmbulo, quem sabe algo desproposital, na realidade, para ilustrar uma discussão mais específica, que vem em tempos recentíssimos ocupando um lugar crescente em esferas filosóficas, artísticas e tecnológicas. Trata-se do debate acerca da capacidade criativa das máquinas, aqui especialmente no que se refere à Literatura. Segundo levantamento recente em áreas especializadas em criatividade computacional, a geração criativa por IA constitui um campo de pesquisa estabelecido, com diferentes abordagens, linguagens e objetivos explorados. Tal riqueza já se verificava, por exemplo, no levantamento (a essa altura, de um longínquo 2017) do Professor da Faculdade de Ciência e Tecnologia da Universidade de Coimbra Hugo Gonçalo Oliveira, em um artigo que promove vasta revisão bibliográfica do estado da arte do tema, intitulado *A Survey on Intelligent Poetry Generation: Languages, Features, Techniques, Reutilisation and Evaluation*, para ficarmos em amostragem ao mesmo tempo reduzida e precisamente significativa.

Essa diversidade, contudo, evidencia também a dificuldade em se estabelecer critérios objetivos de avaliação estética, mesmo entre humanos, o que leva à multiplicação de sistemas poéticos algorítmicos que priorizam a forma ou a surpresa em detrimento da profundidade literária. Como destaca o Professor de Literatura Pedro D'Alte, o desenvolvimento dos chamados *poetry bots* se insere em um movimento tecnológico que busca dotar robôs e algoritmos de uma «habilidade linguística» suficiente para que sua produção textual seja percebida como natural, humana e, em certos casos, literária. Tal tentativa de produzir poesia por algoritmos se ancora, porém, numa performance que frequentemente substitui o valor estético por nada mais do que um mero efeito de reconhecimento: não se pergunta se o texto é poético, mas tão somente se «soaria humano» — e assim precisaríamos ressaltar novamente: é de uma ilusão de espelhamento que se trata o processo (mas não nos antecipemos ainda).

Em um cenário onde *chatbots* como ChatGPT, Bard e Sydney se apresentam como ferramentas capazes de produzir textos literários em prosa ou em verso, surge a dúvida: seria a produção literária dessas supostas inteligências artificiais uma forma legítima de criação ou um simples truque dissimulado? Para investigar essa questão, voltemo-nos ao papel da aleatoriedade na criação literária, o que faremos à luz de já centenários movimentos (um dia vanguardistas) como o Dadaísmo e o Surrealismo e de conceitos borgianos sobre o tema, além da relação que vislumbramos que se podem daí estabelecer com os processos algorítmicos envolvidos na produção de textos pela dita Inteligência Artificial.

O dilema da criatividade

No conto *El inmortal*, o escritor argentino Jorge Luis Borges propõe uma reflexão sobre a produção literária humana ao afirmar: «Homero compuso la Odisea; postulado un plazo infinito, con infinitas circunstancias y cambios, lo imposible es no componer, siquiera una vez, la Odisea».[2] A frase, borgiana por excelência, sugere que, com um tempo e um contexto infinitos, a (re)escrita de obras literárias se tornaria inevitável. Confirma-o em *La Biblioteca de Babel*: «todos los libros, por diversos que sean, constan de elementos iguales: el espacio, el punto, la coma, las veintidós letras del alfabeto», «No hay en la vasta Biblioteca, dos libros idénticos», «a Biblioteca es total y [...] sus anaqueles registran todas las posibles combinaciones de los veintitantos símbolos ortográficos (número, aunque vastísimo, no infinito) o sea todo lo que es dable expresar: en todos los idiomas».[3]

---

[2] Borges 1957, 21.
[3] Borges 1956, 93-94.

O que o autor argentino parece indicar, nesse sentido, não é que o processo de criação se torna uma questão de probabilidade, mais do que de talento ou inspiração. Assim seria se nos fosse dada a Eternidade, esse fundamento que foi tão caro aos escritos filosóficos do criador da Biblioteca e que ele reconheceu que civilizações antigas atribuíram ao divino. Embora o ser humano não disponha de um tempo infinito para criar (então, carece do uso da criatividade), as máquinas, ao contrário, estão cada vez mais próximas dessa fronteira, dispondo quase de um tempo divinamente eterno, dada a velocidade de processamento cada vez maior. A IA, com sua capacidade de acessar e processar uma quantidade incalculável de informações (*elementos iguales, aunque vastísimo, no infinito*) em uma velocidade impressionante — poucos segundos, quando muito — cria assim essa espécie de ilusão de criatividade, aproveitando-se desse traço limitador que o humano guarda em relação à máquina.

Mas recuemos um pouco. O «Magic Realism Bot», lançado em 2015, foi uma das primeiras manifestações de uma IA voltada à criação literária, gerando microcontos com um estilo que, segundo seus idealizadores, estaria a remeter ao Realismo Mágico. No entanto, o que o *bot* fazia de fato era produzir sequências de palavras e frases com pouca ou mesmo nenhuma coerência semântica, geradas a partir de uma mínima estrutura sintático-morfológica pré-estabelecida. Essa produção, por mais divertida ou inusitada que fosse, reclamava profundidade e ostentava uma evidente carestia de efetiva compreensão dos elementos literários. A tentativa de atribuir a esses textos a alcunha de «Realismo Mágico» fazia parte do truque de seus idealizadores na dissimulação do que seria a tal 'inteligência': uma tentativa de tornar o arbitrário palatável ao público-leitor, criando uma expectativa de significado que, do contrário, nunca ou dificilmente se concretizaria.

Se os principais teóricos do conto apontam esse gênero como uma narrativa concentrada, «máquina infalível destinada a cumprir sua missão narrativa com a máxima economia de meios»[4] como queria Julio Cortázar, o microconto, para se realizar como tal, deverá ser a radicalização desse processo, sua máxima condensação, isto é, sua leitura precisaria criar toda uma narrativa a partir de uma explosão de significados,

---

[4] Cortázar 2008, 228.

como o Big Bang criou o Universo. Ora, frases como «Bisexuality has been drinking cocktails all afternoon» ou «A witch secretly leaves her palace at midnight to have sexual intercourse with full communism», produzidas pelo «Magic Realism Bot»,[5] não possuem, na prática, a qualidade explosiva de sentidos que caracteriza famosos microcontos que, a partir de sua pequena dimensão, efetivamente explodem em significados, como o instigante e múltiplo «Cuando despertó, el dinosaurio todavía estaba allí»[6] do guatemalteco Augusto Monterroso, o pungente «For sale: baby shoes, never worn»[7] do romancista estadunidense Ernest Hemingway ou o engenhoso «Machine» do roteirista de HQs também estadunidense Alan Moore «machine. Unexpectedly, I'd invented a time»,[8] que usa a própria linguagem para apresentar o enredo de sua narrativa, em uma metalinguagem criativa de difícil alcance a qualquer IA.

A questão central, então, é: o que o «Magic Realism Bot» fazia não era, de fato, criar uma obra literária genuína. Em vez disso, ele simplesmente organizava palavras e frases de maneira aleatória, sob uma sintaxe mínima e prosaica do tipo sujeito+verbo+complemento+[às vezes] adjunto,[9] em um processo que remonta mais às propostas de Tristan Tzara e ao Dadaísmo do que a qualquer tradição literária estabelecida.

---

[5] Balbi 2021.

[6] Monterroso 1981, 77.

[7] Cox 2008, 42.

[8] Moore 2006.

[9] Hugo Gonçalo Oliveira fora mais preciso em explicar tais procedimentos: «To avoid the generation of poems completely from scratch, most poetry generators take shortcuts and rely on human-created text, usually poems, for inspiration. Different systems exploit the inspiration set differently, generally for the acquisition of useful knowledge or guidelines that will simplify how certain features (e.g. form, syntax or even figurative language) are handled, and also to help modelling the produced poems towards recognisable poetry. This is also reflected on how the inspiration contents are reused in the produced poems. Some systems acquire full lines or fragments from human-created poems, rap lyrics, blog posts, or tweets, and recombine them in new poems, considering features such as metre, rhymes, presence of certain words or semantic similarity. On the one hand, these solutions minimize the issues of dealing with syntax and do not require an underlying generation grammar or template. On the other hand, from the point of view of novelty, they are poor, as lines from known texts can be spotted in the middle of the produced poems. If precautions are not taken, this can even lead to licensing issues. Other systems operate on templates to be filled with new words. Templates can be handcrafted and cover

Estamos, por óbvio, sendo injustos com a comparação. Tzara, em seu famoso manifesto *Pour faire un poème dadaïste*, propunha a criação de textos sem qualquer preocupação com sentido ou coesão, visando exatamente a aleatoriedade e o absurdo. Lembremos seu famoso poema de 1916:

> Prenez un journal.
> Prenez des ciseaux.
> Choisissez dans ce journal un article ayant la longueur que vous
>   comptez donner à votre poème.
> Découpez l'article.
> Découpez ensuite avec soin chacun des mots qui forment cet ar-
>   ticle et mettez-le dans un sac.
> Agitez doucement.
> Sortez ensuite chaque coupure l'une après l'autre dans l'ordre
>   où elles ont quitté le sac.
> Copiez consciencieusement.
> Le poème vous ressemblera.
> Et vous voilà «un écrivain infiniment original et d'une sensibi-
>   lité charmante, encore qu'incomprise du vulgaire».[10]

Da mesma forma, vale recordar o famoso jogo do Surrealismo do *cadavre exquis*, que gerava frases sem lógica aparente. Embora seja de amplo conhecimento público do que se trata esse método de produção da poética surrealista, remetamo-nos à definição teórica fornecida na *Antologia do Cadáver Esquisito*, do escritor português Mário Cesariny:

> Jogo de papel dobrado que consiste em fazer compor uma frase ou um desenho por várias pessoas, sem que nenhuma delas possa aperceber-se da colaboração ou colaborações precedentes. O exemplo, tornado clássico, que deu nome ao jogo, está contido na primeira frase obtida deste modo: *O cadáver-esquisito-beberá-o-vinho-novo*.[11]

---

variations of similes and key phrases from newspapers, or they can be extracted automatically from text. The latter kind may be based on full poems or on single lines, Where some words are replaced by others with similar grammatical features, possibly further constrained on metre, rhyme, POS or semantics. For instance, full poems can have content words stripped. Systems that reuse full fragments may also include an additional step where certain words are replaced, in order to better satisfy the target features», Oliveira 2017, 16.

[10] Tzara 1982, 404.

[11] Cesariny 1989, 95.

Percebamos que o *bot* em análise mais se aproxima desse procedimento (também dadaísta) do que de uma qualquer tentativa de criar um conto com coerência narrativa ou significado profundo, não por convicção artística, como faziam os seguidores do Manifesto de André Breton, mas por pura conveniência e mesmo desonestidade artística e desconhecimento literário (porque atribui uma classificação imprecisa ao classificar como Realismo Mágico algo muito mais próximo de movimentos artísticos anteriores — o que, sem dúvidas, menos importava aos seus desenvolvedores, o artigo de Vasques da Cunha com que iniciamos nossa preleção deixa isso bem evidente).

Curiosamente, em uma etapa posterior, quando mais apta a ser vendida como capaz de produzir textos coerentes, um processo oposto poderá ser aplicado, como descreve D'Alte em seu já referido mas não nomeado artigo de 2020, *Inteligência artificial e poesia*, nesse caso falando dos *poetry bolts*.[12] O autor descreve um experimento em que são apresentados poemas para leitores sob ocultação de autoria, com a solicitação de que se aponte se a produção é robótica ou não, no entanto, são selecionados propositalmente versos humanos nonsenses para a comparação. E questiona:

> O que seria de esperar de um acervo nonsense? Admitindo que um leitor-tipo assume que o absurdo é um traço padronizado da escrita robótica, por que razão os administradores do site insistem na colocação de tais poemas? Poderia supor-se que, dado o exemplo, o computador enganou o humano e é, por isso, criador de poesia? [E conclui] que se pretende que o leitor tome o computador como produtor de poesia à boleia de um exercício altamente condicionado e com um resultado relativamente previsto. Trata-se, assim, de uma publicidade rentável e unidirecional às capacidades dos algoritmos na medida em que não se pergunta: este texto é poético?[13]

D'Alte enfatiza, por conseguinte, o modo como os desenvolvedores de tais tecnologias vendem seus produtos e suas hipotéticas habilidades, superlativando-as, replicando de tal modo essa espécie de efeito de encantamento algorítmico — insistamos na alegoria: o ilusionismo do jogo de espelhos. Para ele,

---

[12] D'Alte 2020.
[13] D'Alte 2020, 169.

esse tipo de proposta, ao evitar qualquer crítica textual real, reduz a leitura a um gesto binário — humano ou não-humano — ocultando o verdadeiro esforço literário e deslocando o valor estético para o engano performático.

Ao compararmos, porém, o «Magic Realism Bot» e mesmo os *poetry bots* de 2020 com o ChatGPT ou outras IAs mais avançadas de uma ou meia década mais tarde, percebemos que a diferença fundamental está na sua capacidade de gerar textos que se aproximam, de maneira sintática e semântica, do que poderíamos considerar como uma obra literária (muitíssimo) mais elaborada. Isso ocorre porque o ChatGPT e seus concorrentes não apenas organizam palavras de forma aleatória em uma frase com sintaxe minimamente coerente, mas escolhem, a partir de um vasto banco de dados, a combinação mais próxima de um estilo, de um tema ou de uma ideia que faça sentido dentro de um contexto linguístico específico a partir de uma gama efetivamente maior de opções. Essa habilidade da IA é o que confere a ela um poder ilusionista (igualmente muitíssimo) mais intenso e eficaz: ela nos faz acreditar que está criando algo genuinamente novo e criativo, enquanto na verdade apenas combina elementos existentes de maneira a gerar um texto que se aproxima daquilo que esperamos de uma produção literária (como, ademais, realiza em diversas outras tarefas). É um pouco mesmo como o truque de um mágico, que nos mostra algo que parece impossível, mas que, ao analisarmos mais de perto, revela-se um processo perfeitamente racional e calculado.

A questão aqui não é apenas técnica, mas também filosófica. Como D'Alte observa ainda, o reconhecimento de um texto como poético ou literário não deveria depender de uma simples categorização binária, mas da capacidade do leitor de mobilizar critérios de apreciação estética, densidade formal e profundidade semântica. Quando essa mediação é abandonada — como no exemplo da classificação automática de poemas em «mais humanos» ou «mais robóticos» — promove-se uma superficialização da leitura e, com ela, uma diluição dos critérios que sustentam a própria ideia da Literatura propriamente dita, em toda a sua complexidade — essa prática tão absolutamente humana, talvez mais do que qualquer outra arte, oriunda de algo tão pouco natural ao ser humano que é o ato de ler e escrever (apesar do seu instinto inegável de contar e ouvir histórias). A inteligência artificial, nesse contexto, se torna menos um produtor estético e mais um simulador de verossimilhanças.

Essa inquietação sobre a natureza da criação literária por meio da IA se conecta diretamente às críticas formuladas por Noam Chomsky, Ian Roberts e Jeffrey Watumull, em artigo publicado no jornal *New York Times*, intitulado *The False Promise of ChatGPT*, em 8 de março de 2023. Segundo os autores, modelos como o ChatGPT são «a lumbering statistical engine for pattern matching, gorging on hundreds of terabytes of data and extrapolating the most likely conversational response or most probable answer to a scientific question»,[14] que operam com base na probabilidade de ocorrência linguística, e não movidos por processos cognitivos autênticos ou explicativos. «The crux of machine learning is description and prediction; it does not posit any causal mechanisms or physical laws»,[15] afirmam os linguistas e cientistas, ressaltando que a inteligência humana é definida justamente pela capacidade de formular explicações — e não apenas repetir padrões.

Ao analisar a diferença entre descrever e explicar, os autores ilustram a limitação das IAs com o clássico exemplo da maçã que cai. Para eles, a verdadeira inteligência se manifesta quando, além de prever a queda, poder-se dizer *por que* isso ocorre — ou seja, formular uma explicação baseada em causalidade, como «a maçã caiu por causa da força da gravidade». Nesse sentido, a mente humana busca criar explicações, enquanto os modelos algorítmicos, como o ChatGPT, se limitam à inferência estatística de correlações. A ausência de uma capacidade explicativa real revela não apenas uma limitação técnica, mas uma barreira ontológica: essas máquinas não pensam, elas apenas processam — processam numa velocidade inimaginável ao ser humano, já o destacamos, mas nada mais do que isso.

Essa crítica se alinha à nossa argumentação de que o ChatGPT e seus irmãos Karamazov agem como ilusionistas. Trata-se, digamos, de «poetas fingidores» mas não no sentido pessoano de um fingimento que sente com a alma, o fingimento poético deliberado de quem finge a dor que sente deverasmente. Antes e até longe disso, referimos um mecanismo

---

[14] Chomsky 2023.
[15] Chomsky 2023.

que apenas simula a criação, disfarçando juntamente sua própria vacuidade criativa com uma superfície de coerência e estilo. Chomsky e seus coautores reforçam essa ideia ao apontar que «ChatGPT and similar programs are, by design, unlimited in what they can "learn" (which is to say, memorize); they are incapable of distinguishing the possible from the impossible».[16] Tal constatação reafirma a crítica à falsa profundidade literária produzida pela IA, que imita significância mas carece de intencionalidade genuína.

Outro ponto de consonância está na discussão sobre a moralidade algorítmica. No artigo de Chomsky, destaca-se o dilema ético da IA ao interagir com temas controversos, como na famosa experiência com a IA Tay da Microsoft em 2016, que rapidamente foi corrompida com discursos de ódio. Segundo os autores, na ausência de uma capacidade de raciocinar a partir de princípios morais, o ChatGPT foi grosseiramente restrito por seus programadores de contribuir com qualquer conteúdo inovador em discussões controversas, ou seja, importantes. Isso revela um paradoxo entre criatividade e responsabilidade: ao buscar restringir moralmente a IA, sacrifica-se sua suposta liberdade criativa; ao permitir que ela explorasse qualquer tema, correr-se-ia o risco de produzir conteúdo ético e intelectualmente irresponsável.

Esse paradoxo se traduz também na crítica mais ampla dos autores à indiferença moral nascida da não-inteligência, uma expressão que ecoa com potência no contexto da dissimulação algorítmica analisada neste nosso ensaio (pensamos que podemos assim o classificar). Quando o ChatGPT responde a questões éticas com neutralidade absoluta, ele não apenas evita o conflito, mas revela a natureza maquinal de sua produção: um gesto de «plagiarism and apathy and obviation»,[17] como diagnosticam os autores do artigo do *TNYT*. Assim, a IA não apenas «imita» (ou busca imitar) a Literatura, mas também «desativa» a complexidade ética e subjetiva que constitui uma criação literária autêntica.

Portanto, se o ChatGPT representa uma forma de fingimento literário — uma simulação que mascara a ausência de intenção estética e reflexão crítica — as análises de Chomsky, Roberts e Watumull aprofundam essa crítica ao mostrarem

---

[16] Chomsky 2023.
[17] Chomsky 2023.

que, mesmo em termos cognitivos e éticos, o que essas máquinas oferecem é um substituto inócuo e falso (fingidor, ilusionista) da inteligência. D'Alte aponta que o engano promovido por certos *poetry bots* é construído com base deliberada (pelos programadores, óbvio está) em «atitudes falaciosas» e «manipulações do leitor», que por sua vez não é convidado a refletir sobre a literariedade do texto, mas apenas acerca da sua aparência autoral. Tal estratégia é reveladora de um movimento maior: o da tentativa de legitimar a produção robótica como arte por meio da estetização e não pela efetiva profundidade estética. Mas Chomsky demonstra que o truque ilusionista da IA literária não se restringe ao plano da estética, mas também à moral e à epistemologia: trata-se de um fingimento total, que não sente, não pensa e não se responsabiliza — apenas reproduz.

E isso guarda intencionalidade outra, decerto.

À guisa de conclusão

Retomemos a ideia do «poeta fingidor», na acepção mais literal do termo, já que ela nos surgiu no meio do caminho como uma pedra drummondiana. A expressão remete à ideia de que a IA não cria de fato, mas sim simula a criação, dissimulando sua ausência de real inspiração ou autenticidade. Essa simulação pode ser mais ou menos convincente, dependendo da sofisticação do modelo, mas nunca alcança a complexidade da criação humana, que envolve fatores como a subjetividade, a experiência pessoal e a reflexão profunda sobre a condição humana.

Se, por um lado, Roland Barthes, em seu famoso ensaio de 1967 *A Morte do Autor*, argumentava que o sentido de um texto é gerado, em última instância, pelo leitor, então poderíamos sugerir que a «morte do autor» também se aplica à figura da máquina. No entanto, essa ideia de «morte» autoral, que Barthes introduziu de maneira provocativa, não implica a ausência total de um criador. Ela apenas desloca a origem da criação para o campo do leitor, que, ao interagir com o texto, constrói seu próprio significado. Ora, no caso da IA, o leitor também desempenha um papel na criação do sentido, fá-lo-á com quaisquer obras, evidentemente, mas o texto em si é produto de um algoritmo, não de um sujeito criativo. Não se pode matar o que nunca vivo esteve.

Pensemos um pouco mais sobre isso para não parecer mero jogo retórico. É nesse mesmo passo que, dois anos mais tarde, Michel Foucault profere a conferência *O que é um autor?*, na qual defende que a marca de um autor está unicamente na singularidade de sua ausência; mas, ora, tal ausência pressupõe um esforço, um movimento (um *gesto*, como interpretará Giorgio Agamben em sua leitura do trabalho do filósofo francês), uma presença anterior portanto, enquanto a máquina é ausente desde sempre e para sempre e, talvez ao contrário, como nos demonstrou D'Alte, esforce-se por simular-se presente. Quiçá seja isso: nas antípodas da visão barthesiana, a marca desumana da produção pseudoliterária generativa resida justamente nesse virtuosismo estético pirotécnico de querer parecer (ilusoriamente) sempre um autor presente, o autor vivo que não é.

E resta uma questão de maior gravidade: se, como ainda ponta Agamben, em sua ausência deliberada, o autor permanece como o instaurador da discursividade, isto é, como o selecionador do que se enuncia e do que se pauta, no ilusionismo da produção generativa, quem exerceria tal função? Não esqueçamos que há sempre um engenheiro programador por trás de cada máquina. Não por acaso figuras poderosas que citávamos no início, referidas no artigo de Martim Vasques da Cunha, tenham nela tanto interesse e considerem a IA a «bomba atômica» do nosso tempo. Aguardemos os próximos cinco anos.

Bibliografia

Agamben, Giorgio. 2007. «O autor como gesto». *Profanações*. São Paulo, Boitempo. 49-57.

Balbi, Clara. 2021. «Robôs já escrevem de poemas e microcontos no Twitter a romances inteiros». *Folha SP*. 8 janeiro 2021. ‹http://www1.folha.uol.com.br/ilustrada/2021/01/robos-ja-escrevem-de-poemas-e-microcontos-no-twitter-a-romances-inteiros.shtml›

Barthes, Roland. 2004. «A morte do autor». *O rumor da língua*. São Paulo, Martins Fontes.

Borges, Jorge Luis. 1957. *El Aleph*. Buenos Aires, Emecé.

Borges, Jorge Luis. 1956. *Ficciones*. Buenos Aires, Emecé.

Cesariny, Mário. 1989. *Antologia do cadáver esquisito*. Lisboa, Assírio & Alvim.

Chomsky, Noam, Ian Roberts e Jeffrey Watumull. 2023. «The False Promise of ChatGPT». *The New York Times*, 8 março 2023. ‹http://www.nytimes.com/2023/03/08/opinion/noam-chomsky-chatgpt-ai.html›

Cox, Alisa. 2008. *The Short Story*. Newcastle, Cambridge Scholars Publishing.

Cunha, Martim Vasques da. 2025. «Quem são os intelectuais bilionários que preparam a ruptura apocalíptica de Trump». *Folha SP*. 19 abril 2025. ‹http://www1.folha.uol.com.br/ilustrissima/2025/04/quem-sao-os-intelectuais-bilionarios-que-preparam-a-ruptura-apocaliptica-de-trump.shtml›

D'Alte, Pedro. 2020. «Inteligência artificial e poesia: Uma reflexão sobre o caso dos "Poetry bots"». *Revista 2i: Estudos de Identidade e Intermedialidade* 2.2: 165-77.

Foucault, Michel. 2006. *O que é um autor?* Lisboa, Nova Vega.

Monterroso, Augusto. 1981. *Obras Completas (y otros cuentos)*. Barcelona-Caracas, Seix Barral.

Moore, Alan. 2006. «Machine». *Wired*. 1 novembro 2006. ‹http://wired.com/2006/11/very-short-stories›

Nolan, Christopher. 2010. *Inception*. Burbank, Warner Brothers Pictures.

Oliveira, Hugo Gonçalo. 2017. «A Survey on Intelligent Poetry Generation: Languages, Features, Techniques, Reutilisation and Evaluation». *Proceedings of the 10th International Conference on Natural Language Generation*. Santiago de Compostela, 11-20. ‹http://aclanthology.org/W17-35.pdf›

Tzara, Tristan. 1982. «Pour faire un poème dadaïste». Em *Poésie française: anthologie critique*. Marie-Louise Astre, Françoise Colmez e Philippe Soupault, eds. Paris, Bordas. 404.

«La velocidad de lo real»: The Literary Apocalypse

Sandra Sousa

The memories remain vivid, even five years later. Boston, March 2020. I was attending the Northeast Modern Language Association Conference, surrounded by colleagues from across the country and abroad. For the first time, the usual warmth of hugs and handshakes was conspicuously absent. At first, I didn't think much of it. There were vague murmurs about some virus no one seemed to fully understand, but I brushed them off. Maybe it was my habit of staying less informed than I should, or maybe I just thought people were overreacting.

The unease began to grow as I traveled from Boston to Milwaukee. By the time I was sitting in the airport, waiting for my flight back home, the whispers had turned into something louder. Then, the chime on my phone: an announcement from the university that Spring Break would be extended to give faculty time to move classes online. Reality began to sink in, though in fragments. A mother seated nearby wiped down her daughter's tray table with meticulous care, casting cautious glances my way as though I carried something contagious. What was happening?

This story, with its own variations, belongs to nearly everyone on the planet. It was the beginning of a shared global upheaval. At the time, as the world was being boxed into isolation — no travel, minimal human contact, the routines of «normal life» upended — I felt something unexpected: peace. While many around me were spiraling into fear, breaking safety rules, and grasping for control, I felt a kind of relief. Maybe, just maybe, this was the moment humanity would pause long enough to learn something vital. Perhaps this forced slowing down would unravel the frantic pace of life dictated by our exploitative, capitalist systems. Maybe, in this collective stillness, people would reflect deeply on what we were doing wrong — both as individuals and societies — and choose something better. Maybe we would reach for the best in ourselves and finally build a world that was more just, more compassionate, and more equitable for all.

The post-pandemic reality, however, painted a different picture as the doors began to open once again. Instead of embracing a newfound awareness or a collective commitment to change, it felt as though much of the world rushed to restore the very systems and routines that had been paused. The hope that humanity might reflect deeply on its priorities, that we might emerge with a clearer vision for justice, equity, and balance, seemed to dissipate as quickly as lockdowns lifted. For a fleeting moment, it had felt possible; like we stood on the brink of transformation. Yet, as normalcy returned, so too did the old habits: the unrelenting pace, the widening inequalities, the frenetic scramble to «catch up» on lost time. What I had imagined as a turning point now felt more like a missed opportunity, a chance to rebuild squandered in the rush to return to what was, rather than reimagine what could be.

Three years later, a new kind of virus seemed to be embedding itself in our lives, one that most people appear to be cheering for rather than questioning. Unlike its biological predecessor, no lockdown will stop this one if we fail to act with care and foresight. In 2023, my university hosted a massive conference titled *Teaching & Learning with AI.* Among the 500 attendees, only a handful of colleagues, myself included, seemed uneasy about the promises of AI for the future of education. I delivered what many might call a «doom and gloom» talk, and by the end, the room was eerily silent. A couple of attendees approached me afterward, not to disagree, but to quietly thank me for saying aloud what others were hesitant to voice. Still, the overwhelming response was one of uncritical enthusiasm. The message of the conference was clear: AI is not just here to stay but to dominate, and we, as educators, must embrace it. The initial resistance I felt among my peers soon dissolved under administrative pressures urging us to incorporate AI into our work as teachers and researchers.

The fragile sense of peace and reflection I had experienced during the pandemic now feels like a distant memory. Once again, I find myself wondering what is wrong with my fellow humans. Do I have an inherent tendency toward pessimism? Or are my concerns rooted in something real? What will happen to my field of literature, for instance, if we wholeheartedly embrace an intelligence capable of writing essays, articles, and even novels?

The more I research, the more this fear solidifies, not as a rejection of innovation but as a recognition of the risks AI poses to the arts, the humanities, and our collective humanity. This is not a fear conjured by imagination but grounded in reality. To explore these reflections further — what I call the «literary apocalypse», borrowing a phrase from Colombian author Mario Mendoza — I will draw on the works of writers from disparate and often marginalized parts of the world: Mia Couto from Mozambique and Mario Mendoza from Colombia. In tandem with their perspectives, I will turn to Mo Gawdat, a former chief business officer at Google X, and Yuval Noah Harari, an historian and philosopher who has grappled with the ethical dilemmas of technology. Together, these voices will, hopefully, help navigate the uneasy intersection of AI, literature, and our future as human societies.

In 2021, renowned Mozambican writer Mia Couto published a collection of short stories titled *O Caçador de Elefantes Invisíveis* (*The Invisible Elephant Hunter*). The collection opens with three stories centered on the Covid-19 pandemic, reflecting on its profound impact on humanity. The very first story, *Um Gentil Ladrão* (*A Gentle Thief*), sets the tone for the book, offering a poetic and deeply symbolic exploration of the pandemic's disruption to our lives and its moral and existential consequences. *Um Gentil Ladrão* compels the reader to meditate on isolation, systemic neglect, and human fragility, using the Covid-19 pandemic as a backdrop to examine broader societal issues. The narrative begins with an elderly man living in extreme isolation, far removed from the conveniences of urban life. The knock on his door — a rare event — introduces a masked health worker whose clumsy actions set the tone for the story's mix of satire and poignancy. From the outset, Couto establishes the narrator's estrangement not only from people but from the very systems meant to safeguard human life. As the narrator remarks, «A minha falecida mulher dizia que a culpa era nossa porque escolhemos viver longe dos lugares onde há hospitais. Ela, coitada, não sabia que era o inverso: os hospitais é que se instalam longe dos pobres».[1] His reflection

---

[1] Couto 2022, 15. «My late wife used to say it was our fault for choosing to live far from places with hospitals. Poor woman, she didn't know it was the opposite: hospitals are the ones that place themselves far from the poor» (translation mine).

critiques the systemic inequalities that have long left marginalized communities to endure crises alone. For the narrator, the pandemic merely reveals what has always been present: a world where care and resources are distributed unevenly, and where the poor are left to fend for themselves.

The encounter between the narrator and the health worker becomes a microcosm of the pandemic's tensions. The worker's overzealous adherence to protocols — keeping a three-meter distance, wearing plastic bags on his shoes, and wielding a thermometer resembling a gun — highlights the absurdity and alienation engendered by pandemic measures. As the narrator describes, «Retira da bolsa uma pistola. Aponta-a na minha direção. É estranha aquela arma: de plástico branco, emitindo um raio de luz verde».[2] This scene underscores the fear and distrust that can arise when technological solutions are implemented without genuine human connection. Yet, the narrator's deeper insight lies in his recognition that the true pandemic is not the virus itself but a more pervasive disease: «Então, me vem à cabeça o nome da doença de que fala o visitante. Conheço bem essa doença. Chama-se indiferença. Era preciso um hospital do tamanho do mundo para tratar essa epidemia».[3] By framing indifference as the real affliction, Couto moves beyond the immediate context of Covid-19 to critique the apathy and systemic failures that have long shaped the lives of the disenfranchised.

Despite its critique of societal neglect, *Um Gentil Ladrão* ultimately finds a glimmer of hope in human connection. The narrator's defiant embrace of the health worker, despite the latter's resistance, symbolizes a yearning for solidarity, even in the face of fear and alienation. While the health worker departs hurriedly, shedding his clothes «como se despisse as vestes da própria peste»,[4] his parting gesture leaves a lasting

---

[2] Couto 2022, 13. «He pulls a pistol from his bag. Points it at me. It's a strange weapon: white plastic, emitting a green ray of light» (translation mine).

[3] Couto 2022, 16. «Then the name of the disease the visitor speaks of comes to mind. I know this disease well. It's called indifference. We would need a hospital the size of the world to treat this epidemic» (translation mine).

[4] Couto 2022, 16. «as if he were shedding the very garments of the plague itself» (translation mine).

impression. He leaves behind soap, hand sanitizer, and hygiene instructions; items meant to protect against the virus but also symbolic of a fleeting attempt to care for the narrator. In a moment of humor and resignation, the narrator reflects, «Para a semana, quando ele voltar, vou deixar que roube a velha televisão que tenho no quarto».[5] This final line captures the story's delicate balance between despair and hope, showing that even in a world plagued by indifference, small acts of connection and care can offer a reprieve from isolation. Through this intimate and layered narrative, Couto transforms the pandemic into an allegory for humanity's enduring struggle against neglect, revealing both its vulnerabilities and its resilience.

The intricate interplay between solitude and connection that Mia Couto masterfully crafts in *Um Gentil Ladrão* finds a poignant continuation in *A Imortal Quarentena*. While the former highlights the external visitor's role in rekindling humanity amidst indifference, the latter turns inward, portraying the psychological isolation of a solitary writer. Here, the protagonist, Bernardo, is a writer trapped in his own thoughts and fears, exacerbated by the pandemic's isolating effects. The opening description, «Bernardo acorda tarde, deprimido por ter acordado, angustiado por ser sempre tarde»,[6] immediately introduces us to a man caught in the grip of existential despair. Like the old man in *Um Gentil Ladrão*, Bernardo struggles with the weight of his own isolation, though it is self-imposed rather than imposed by external forces. The pandemic, for him, becomes a backdrop for his ongoing internal conflict, as reflected in his dismissal of ordinary tasks like washing dishes or cleaning the house: «Custa-lhe desperdiçar a sua criatividade em tão rotineiras tarefas».[7] Here, Couto draws a parallel to the first story's focus on existential reflection, but this time, it is framed through a writer's lens, where the internal void and the outside world seem equally distant.

Despite his overwhelming feelings of detachment, Bernardo's emotional journey takes an unexpected turn when he

---

[5] Couto 2022, 16. «Next week, when he returns, I'll let him steal the old television I have in the bedroom» (translation mine).

[6] Couto 2022, 17. «Bernardo wakes up late, depressed for having woken up, anguished for always being late» (translation mine).

[7] Couto 2022, 17-18. «He resents wasting his creativity on such routine tasks» (translation mine).

finds himself confronting the reality of his own social isolation and privilege. He contemplates his relationship with his domestic worker, Dona Esperança, recognizing the boundaries he has imposed in his life: «O vírus é cego, mas a quarentena tem as suas hierarquias sociais».[8] This remark exposes Bernardo's growing awareness of the inequalities that have always existed around him, much like the elderly man in «Um Gentil Ladrão» who recognizes the disparity between the privileged and the marginalized. As Dona Esperança provides him with comfort, he begins to recognize the humanity in the most unlikely places, symbolized by her simple act of singing to him: «Dona Esperança canta para que ele adormeça».[9] The idea of the «musa inspiradora» or «inspirational muse» finally manifests in the form of Esperança, who, despite her humble circumstances, offers Bernardo not just physical care, but the emotional support that rekindles his creativity. In this quiet, humble interaction, Couto suggests that true inspiration often comes from the most unexpected sources; those who have been pushed to the margins of society but retain the deepest forms of humanity.

In *A Imortal Quarentena*, the writer's journey from despair to a tentative reconciliation with the outside world mirrors the thematic concerns of *Um Gentil Ladrão*, wherein the protagonist finds solace in an unexpected human connection. Both stories reflect the broader societal fractures exposed by the pandemic, with Couto using his characters' personal journeys to explore themes of isolation, existential reflection, and the slow, sometimes painful, recognition of social inequalities. Through these two stories, Couto's writing compels us to consider how crises like the pandemic illuminate not only the fragility of our bodies but also the fragility of our social bonds.

The last sentence of *A Imortal Quarentena*, «Pela primeira vez, depois do início da quarentena, Bernardo acorda, abre as cortinas, contempla a rua e recusa estar perante a derradeira

---

[8] Couto 2022, 19. «The virus is blind, but the quarantine has its social hierarchies» (translation mine).

[9] Couto 2022, 20. «Dona Esperança sings to him so that he can sleep» (translation mine).

versão da realidade»,[10] is deeply significant. Bernardo's refusal to accept the final version of reality («a derradeira
versão da realidade») signifies his unwillingness to accept the
external world as it is: bleak, confined, and overwhelming. His
act of opening the curtains, often a symbol of hope, renewal,
or new beginnings, also evokes the possibility of seeing the
world in a different light, one that does not necessarily con-
form to the limitations imposed by the pandemic or the societal constraints it accentuates. Thus, then last line of the story
can be seen as an invitation to consider the blurred boundaries between reality and fantasy. It implies that although Bernardo has lived through the stark and often painful realities of
the pandemic, he is still capable of resisting and reimagining a
reality that is more flexible, creative, and perhaps even fantas-
tical. His refusal indicates that the human spirit is always capable of creating alternatives to the world as it is presented to
us, especially in times of crisis. In the context of literature, this
rejection of «the final version of reality» highlights the role of
imagination and narrative in constructing new possibilities,
even when faced with the most difficult of circumstances.

In the story *O Caçador de Elefantes Invisíveis*, Couto continues to explore the pervasive, intangible forces that define our
existence. The story expands Bernardo's existential contem-
plation by addressing the larger, more universal impact of invisible forces. The story's title itself — *The Hunter of Invisible
Elephants* — serves as a metaphor for an invisible, overwhelming presence that cannot be grasped or fully understood, yet
profoundly shapes lives. As the epigraph states, «O invencível
foi derrubado pelo invisível»,[11] it reflects the paradox of how
a force — like the pandemic — can upend even the strongest
structures, revealing the fragility of our understanding of the
world. In this context, the invincible refers to *Homo Sapiens*
and the ego that has often led humanity to believe in its domi-
nance over nature. This sense of invincibility, rooted in human
hubris, is challenged by the invisible forces that, though intangible, possess the power to disrupt and alter the course of life.

---

[10] Couto 2022, 21. «For the first time, after the beginning of the quarantine, Bernardo wakes up, opens the curtains, looks out at the street, and
refuses to face the final version of reality» (translation mine).

[11] Couto 2022, 23. «The invincible was brought down by the invisible»
(translation mine).

The narrative calls attention to how the belief in human invincibility, deeply embedded in the collective ego, can be shattered by forces beyond human comprehension. Just as society once believed itself impervious to a global health crisis, the pandemic exposed the vulnerabilities hidden beneath the surface of human arrogance. Through this lens, Couto invites readers to reflect on how life is often shaped by factors beyond human control, urging a humbler understanding of our place in the world. The story, much like the epigraph, serves as a reminder that even the so-called «invincible» can be defeated by the invisible, urging us to rethink our assumptions about power, control, and the fragility of human existence. Couto's stories are not merely about a virus but also about the societal and personal transformations it precipitates, leaving readers to ponder the thefts, both tangible and intangible, that such a global catastrophe enacts on our lives. His stories reveal that the pandemic is not only a biological phenomenon but also a metaphor for deeper human vulnerabilities and systemic failures. Like the virus that silently spreads through Couto's narratives, there is another «virus» at work in our societies — equally invisible, equally insidious — one that dismantles the illusion of power and control that humanity has long taken for granted. Yuval Noah Harari addresses this troubling reality in the opening of his recent book, *Nexus: A Brief History of Information Networks from the Stone Age to AI*:

> Over the last 100,00 years, we Sapiens have certainly accumulated enormous power. Just listing our discoveries, inventions, and conquests would fill volumes. But power isn't wisdom, and after 100,000 years of discoveries, inventions, and conquests humanity has pushed itself into an existential crisis. We are on the verge of ecological collapse, caused by the misuse of our own power. We are also busy creating new technologies like artificial intelligence (AI) that have the potential to escape our control and enslave or annihilate us. Yet instead of our species uniting to deal with these existential challenges, international tensions are rising, global cooperation is becoming more difficult, countries are stockpiling doomsday weapons, and a new world war does not seem impossible.[12]

Harari's words resonate with Couto's epigraph, reflecting the paradox of modern existence. Humanity — *Homo sapiens*

---

[12] Harari 2024, xi.

— has built an empire of extraordinary discoveries and technological advancements, yet this very power is leading us to the brink of collapse. The pandemic, as depicted in Couto's fiction, serves as a powerful reminder of our fragility and the limits of our control, challenging the ego-driven belief that we are masters of the world. Through the juxtaposition of Harari's warning and Couto's allegorical storytelling, a broader reflection emerges: humanity's greatest threat may not come from an external force but from within—our own unchecked hubris and inability to cooperate in the face of invisible dangers.

Harari contends that even though «nobody disputes that humans today have a lot more information and power than in the Stone Age, it is far from certain that we understand ourselves and our role in the universe much better».[13] This idea strikes at the heart of the modern condition, where technological prowess far outpaces our wisdom and ethical maturity. Humanity stands as the invincible, confident in its triumph over nature, yet it remains vulnerable to forces it has unleashed, from environmental collapse to artificial intelligence. Harari warns that «we have already driven the earth's climate out of balance and have summoned billions of enchanted brooms, drones, chatbots, and other algorithms spirits that may escape our control and unleash a flood of unintended consequences».[14] His invocation of mythical parallels like «The Sorcerer's Apprentice» underscores the inherent dangers of creating tools we cannot fully govern, echoing Couto's allegories where an invisible «virus» — literal or metaphorical — reveals our lack of control over the forces we set in motion.

In this context, the future of literature becomes a critical site of reflection. The modern era, as Harari points out, faces not only environmental and technological crises but also a crisis of informational narratives. He argues that while humanity has unprecedented access to information, this very abundance has created confusion, misinformation, and competing narratives that erode our ability to discern truth. The digital age, with its algorithms and artificial intelligence, amplifies these crises by prioritizing engagement over accuracy, fostering delusional networks that can manipulate societies and undermine critical thinking. In this way, the struggle is not just

---

[13] Harari 2024, xi.
[14] Harari 2024, xiii.

against external challenges but also against the stories we tell ourselves and the fictions we choose to believe: «The tendency to create powerful things with unintended consequences started not with the invention of the steam engine or AI but with the invention of religion. Prophets and theologians have summoned powerful spirits that were supposed to bring love and joy but occasionally ended up flooding the world with blood».[15] Similarly, literature has long served as a tool for questioning and reimagining the narratives that shape humanity. Yet, as Harari further cautions, «some new totalitarian regime may well succeed where Hitler and Stalin failed, creating an all-powerful network that could prevent future generations from even attempting to expose its lies and fictions».[16] The growing influence of artificial intelligence and algorithms in disseminating information poses a direct challenge to literature's role as a beacon of truth and critical thinking. As Harari states, «while fables like the myth of Phaeton and "The Sorcerer's Apprentice" present an overly pessimistic view of individual human psychology, the naïve view of information disseminates an overly optimistic view of large-scale human networks».[17] Literature, therefore, emerges not merely as a cultural artifact but as a necessary form of resistance, a means of exposing the fictions we live by and offering alternative ways of imagining the world.

Couto's allegorical storytelling and Harari's philosophical warnings intersect at a critical juncture: the tension between humanity's ego-driven pursuit of invincibility and its struggle to confront invisible forces, be they pandemics, algorithms, or ideologies. If, as Harari suggests, the solutions do not lie in fables or divine interventions, then literature must take on the responsibility of exposing the unintended consequences of our actions. It is within this tension — between power and wisdom, control and chaos — that literature can illuminate paths forward, compelling us to reflect on the invisible forces shaping our present and our future. Harari's concerns about humanity's informational and technological crises resonate deeply when viewed through the lens of literature's uncertain future in the age of artificial intelligence. AI, like a silent contagion, spreads inexorably, infiltrating creative spaces and replacing

---

[15] Harari 2024, xiii.
[16] Harari 2024, xiv.
[17] Harari 2024, xv.

human imagination with soulless, mechanical narratives. As AI-generated literature proliferates, it raises existential questions about the very essence of storytelling: Will machines, capable of producing vast quantities of text at unprecedented speed, outpace and overshadow human creativity? If storytelling becomes a matter of algorithms, will literature lose its soul, its warmth, struggle, and vulnerability? These human qualities, shaped by lived experience, imperfection, and emotion, cannot be programmed into machines.

This moment can be seen as a *Literary Apocalypse*, where literature risks being reduced to efficient yet hollow productions devoid of the human touch that makes stories resonate across generations. The act of storytelling has always been more than mere language; it carries the weight of humanity's joys, sorrows, and contradictions. AI may master syntax and structure, but it lacks the ability to convey the whispers of doubt, the unexpected turns of thought, or the ineffable beauty found in human imperfection. Literature's heartbeat lies in its vulnerability, its capacity to reveal, question, and connect on a deeply human level.

In this collision of pandemics — both the biological and the artificial — literature finds itself at a reckoning. Just as the Covid-19 pandemic exposed the fragility of human life and systems, the rise of AI reveals the fragility of human creativity in a world increasingly dominated by machines. However, this is not merely a crisis; it is also a challenge, a call for writers, readers, and thinkers to defend the irreplaceable human essence in art. Like Couto's allegories of invisible forces bringing down the seemingly invincible, this technological «virus» forces us to reexamine our role as storytellers and guardians of meaning, lest we allow literature to drift into artificial, lifeless words.

It is here that Mario Mendoza's *Bitácora del Naufragio* (*Diary of a Shipwreck*) offers a striking reflection on the rapid acceleration of our present and the ways in which reality has outpaced fiction. In the chapter *La velocidad de lo real* (*The Speed of the Real*), Mendoza grapples with a disconcerting realization: «¿En qué momento nos habíamos convertido en personajes de nuestros propios libros? ¿En qué momento la literatura de anticipación se nos estaba volviendo puro realismo?»[18] What once belonged to the realm of speculative fiction — pandemics, dystopias, societal collapse — has erupted into our lived experience with unsettling speed, blurring the lines between imagined futures and present realities. Mendoza's accounts of writing graphic novels (*Los híbridos* [*The Hybrids*], *Morgellons*, and *Homo Capensis*), in which characters navigate pandemics eerily similar to Covid-19, reveal the unsettling truth that literature is no longer anticipating disaster but recording it in real time.

This «speed of the real» that Mendoza describes is, at its core, literature's existential dilemma in the 21st century. Fiction, once a space for imaginative speculation, now struggles to keep pace with a world accelerating toward entropy, chaos, and collapse. As Mendoza warns, «Nuestra única opción es avanzar valientemente sabiendo que hemos cruzado el punto de no retorno, y que en esta curva por la que estamos viajando solo nos espera la entropía, el caos y el horror».[19] Unlike the utopian progress we once envisioned, Mendoza argues that our trajectory has veered toward an irreversible dystopia.

Yet, even in the face of this collapse, *Bitácora del Naufragio* provides a crucial insight: the role of literature is not to offer false hope or nostalgic comfort but to confront the abyss with unflinching honesty. Like the works of Asimov, Bradbury, Huxley, and Orwell, literature must bear witness to collapse and give it a name. Mendoza challenges writers and intellectuals to abandon the «nostalgic tone» of recovery and ascendancy, urging them instead to confront what he calls «lo irremediable».[20] In doing so, literature fulfills its most essential function, not as a tool of escapism but as a record of human resilience in the face of unavoidable decline.

In this way, Mendoza's *Bitácora del Naufragio* offers a sobering response to the threats posed by AI and other invisible forces eroding human creativity. If the «literary apocalypse»

---

[18] Mendoza 2021, 210. «At what point had we become characters in our own books? At what point had speculative literature turned into pure realism?» (translation mine).

[19] Mendoza 2021, 215. «Our only option is to move forward bravely, knowing that we have crossed the point of no return, and that on this curve we are traveling, only entropy, chaos, and horror await us» (translation mine).

[20] Mendoza 2021, 214.

looms, it is because we failed to recognize the speed at which the real overtakes the imagined, leaving us unprepared to grapple with its consequences. Literature's challenge, then, is to adapt, not by relinquishing the human touch but by doubling down on it. As Mendoza reminds us, the future may be irreversible, but storytelling remains our lifeline: a way to map the shipwreck, confront entropy, and remind ourselves of what it means to be human in the midst of chaos:

Nuestra única opción es avanzar valientemente sabiendo que hemos cruzado el punto de no retorno, y que en esta curva por la que estamos viajando solo nos espera la entropía, el caos y el horror.[21]

Mendoza's words underscore the inevitability of our situation, yet also highlight the crucial role of literature in navigating this uncertain path, keeping us anchored in our humanity as we face the forces of entropy, chaos, and horror. As Mia Couto also wisely observes, «Somos os mais competentes carcereiros de nós mesmos»,[22] we often become prisoners of our own fears, limitations, and beliefs.

This, indeed, is the true task of literature in the age of pandemics — biological, technological, and existential: to narrate the collapse, not with despair but with insight, courage, and conviction, pushing us to question the limits of our control, and to break free from the self-imposed prisons of our own making. Harari echoes this sentiment, cautioning that «AI is the first technology in history that can make decisions and create new ideas by itself».[23] The implications of this are staggering. As AI takes on an increasingly central role in shaping our future, it becomes crucial for us to retain the wisdom and humanity that storytelling offers. In a world where algorithms and artificial intelligence may increasingly dictate our decisions, literature, much like the work of Mendoza and Couto, serves as a critical counterbalance. It forces us to confront the invisible forces, both biological and technological, that threaten our existence, ensuring that the essence of what it means to be human does not vanish into the void of artificial narratives. Literature, then, is not just a record of our past but a vital tool for grappling with the uncertainties of the future,

---

[21] Mendoza 2021, 215.
[22] Couto 2022, 136. «We are the most competent jailers of ourselves». [my translation]
[23] Harari 2024, xxii.

helping us navigate the ever-accelerating pace of change while reminding us of the complexities of human emotion, struggle, and resilience

Bibliography

Couto, Mia. 2022. *O Caçador de Elefantes Invisíveis*. Lisboa, Caminho.

Harari, Yuval Noah. 2024. *Nexus*. New York, Random House.

Mendoza, Mario. 2021. *Bitácora del Naufragio*. Bogotá, Editorial Planeta.

EFL Listening in the Age of AI: Student Perceptions, Topic Familiarity, and Implications for OER Development[*]

Mingyu Sun

1. Introduction

Artificial intelligence is increasingly transforming the landscape of language learning by enabling the creation of personalized, scalable, and adaptable content. With the advancement of generative AI technologies, particularly large language models such as ChatGPT-4 (www.chatgpt.com), generation of naturalistic dialogues and listening materials tailored to specific learner profiles has become significantly more efficient. Tasks that were once time-consuming and labor-intensive can now be streamlined, opening new possibilities for instructional design. These innovations hold promise for the enhancement of Open Educational Resources (OER), enabling faster content production, greater customization, and broader inclusivity.

Listening comprehension is especially critical for English as a Foreign Language (EFL) learners, as it underpins vocabulary development, oral fluency, and overall communicative competence. A wide range of listening materials is already available through platforms such as English Listening Lesson Library Online (ELLLO), Randall's ESL Cyber Listening Lab, Voice of America Learning English, and TED-Ed and TED Talks, many of which incorporate OER elements. Repositories like OER Commons and MERLOT (Multimedia Educational Resources for Learning and Online Teaching) also provide access to listening-focused educational content.

[*] The author would like to thank Professor Jiankai Yu from Henan University, China for his invaluable support in administering all in-class listening comprehension exams, collecting data, and facilitating communication with his students regarding the study. Special thanks are also extended to his students for their participation and engagement throughout the research process.

At the Language Open Education Resource (OER) Conference hosted by the University of Kansas (https://olrc. ku.edu/oer), language educators showcased diverse approaches to developing listening materials for various proficiency levels, including individual lessons, complete courses, and serialized curricula. Notably, many participants reported incorporating AI tools into their material development processes.

Building on these developments, the present study investigates how college-level Chinese EFL students perceive and comprehend AI-generated versus human-created listening materials. By examining the role of topic familiarity, the study contributes to a deeper understanding of learner comprehension and offers insight into how AI tools can be responsibly leveraged to develop listening-focused OER that are both pedagogically effective and culturally sensitive.

2. Literature Review

2.1. Brief background on the rise of AI in language education

Language educators have witnessed a rapid rise in the use of artificial intelligence (AI) across various aspects of language education. Early tools such as Google Translate and Grammarly introduced AI-powered support for translation and writing assistance. In recent years, many language learning applications that originally lacked AI features have evolved to incorporate them. For example, Duolingo now uses AI algorithms to offer personalized learning paths; Quizlet includes an AI tutor bot; and Elsa Speak leverages automated speech recognition to provide real-time corrective feedback, helping learners improve pronunciation and fluency. Assessment technologies have also adopted AI, with systems like ETS SpeechRater enabling scalable, consistent, and efficient evaluation of spoken language proficiency.

In addition, dozens of AI-based quiz and content generators are now capable of creating customized language exercises, dialogues, and reading passages tailored to specific learner levels and interests. The emergence of large language models (LLMs) enabled technology to make content creation extremely accessible and affordable. Building on these developments, the present study uses Google NotebookLM to generate listening materials on various topics. Google NotebookLM is an AI-powered research and note-taking tool developed by Google Labs, designed to help users synthesize, summarize, and interact with their own documents using advanced language models. Unlike some chatbots, NotebookLM allows users to upload source files such as PDFs, Google Docs or Google Slides, websites and other media such as audio files. The tool then generates summaries, key points, and Q&As based on the inputs. Introduced in September 2024, the tool can also generate engaging, podcast-like audios featuring AI-generated hosts.

2.2. Importance of listening comprehension in EFL learning

Listening comprehension is crucial for developing communicative competence in English as a Foreign Language (EFL) learners. As one of the four core skills, along with speaking, reading, and writing, it often serves as the main source of language input, especially in the early learning stages. However, it typically receives less emphasis in formal instruction, particularly in Chinese English education, where limited class time, practice opportunities, and access to audio materials have led to weaker listening and speaking skills compared to reading and writing.

From a language acquisition perspective, listening is foundational. Research grounded in Krashen's Input Hypothesis emphasizes the need for abundant and comprehensible input for language development, and listening is the main vehicle for that input in most naturalistic learning environments. Through repeated exposure to authentic spoken language, learners internalize vocabulary, develop a sense of grammatical structures, and improve pronunciation. Listening provides the linguistic raw material that supports and shapes productive language use.

Effective listening is critical in both academic and real-world contexts, enabling learners to follow lectures, engage in conversations, and interpret meaning. Without strong listening skills, learners risk missing key information or disengaging from communication.

Listening is also cognitively demanding. Learners must decode sounds, recognize vocabulary, and interpret grammar in real time, often while dealing with accents, noise, or unfamiliar content. A lack of background knowledge can further hinder understanding. Given its central role, pedagogical approaches in EFL should prioritize listening practice that is both frequent and varied. Exposure to a range of speech types, accents, and topics, including both familiar and unfamiliar content, can enhance learners' ability to process diverse linguistic input. Additionally, incorporating materials that are perceived as authentic, engaging, and appropriately challenging is essential for maintaining learner motivation and fostering listening development.

Listening comprehension poses challenges for Chinese EFL learners, making it a critical focus in language instruction. Research has shown that listening-specific vocabulary knowledge, especially academic and topic-related vocabulary, is a key predictor of comprehension success among Chinese university students. Cultural and topic familiarity also play a central role in facilitating listening comprehension. Studies have found that Chinese students demonstrate significantly better understanding when listening tasks are culturally relevant or thematically familiar.

Moreover, the traditional language learning environment in China often lacks exposure to authentic spoken English, which can further hinder learners' ability to process natural speech and diverse accents. Recent instructional innovations, such as flipped classroom approaches, have shown promise in improving listening outcomes and lowering anxiety by giving students more time and support to engage with input-rich materials before class . These findings underscore the importance of designing listening materials that are both linguistically and culturally accessible and suggest that topic familiarity, particularly when paired with effective instructional tools, can significantly enhance EFL learners' listening comprehension in the Chinese context.

2.3. The growing role of OER in resource design and distribution

Open Educational Resources (OER) has become an essential part of the global push toward more accessible, equitable, and flexible education. In the context of language learning, OER offers instructors the ability to freely adapt and share teaching materials that suit specific learner needs, especially in under-resourced or multilingual settings. Recently, the rise of artificial intelligence has significantly accelerated the creation, adaptation, and distribution of OER. AI tools such as generative text models (e.g., ChatGPT), speech-to-text processors (e.g., Whisper), and image generators (e.g., Adobe Firefly) are now commonly used to develop language learning resources, including listening scripts, assessments, and multimedia content.

In addition to improving efficiency, AI-enhanced OER is advancing accessibility through features like automated captioning, voice synthesis, and adaptive visual design aligned with Universal Design for Learning (UDL) principles. However, these developments also raise important questions about authenticity, content bias, and learner trust, especially when resources are generated without human oversight or cultural relevance. As AI tools become more prevalent in OER development, it is critical to understand how learners interact with and perceive these materials.

2.4. OER Targeting EFL Listening Comprehension

Recent research highlights several promising Open Educational Resources (OER) initiatives specifically designed to improve EFL learners' listening comprehension. For example, Chen et al. explored the use of OER videos such as TED Talks in Taiwanese university classrooms and found that they enhanced students' communicative strategies and multicultural competence. Mah and Han evaluated the Self-Assessed Online Listening Tests (SAOLT) platform, an open-access resource offering listening exercises for ESL learners. Educators reported strong perceptions of its ease of use and effectiveness in teaching listening skills. Furthermore, an OER program combining online extensive reading and listening modules demonstrated significant improvements in adult EFL learners' vocabulary, listening comprehension, and motivation. More recently, AI-enhanced podcast activities deployed via an open platform showed statistically significant gains in ESP students' listening comprehension compared to traditional methods.

2.5. Gaps and Rationale for the Current Study

Despite the growing integration of artificial intelligence in language education, there remains limited empirical research comparing AI-generated and human-created listening materials in English as a Foreign Language (EFL) context. Few studies have systematically examined how learners comprehend and engage with AI-generated content in comparison to traditional, educator-authored resources. Furthermore, the role of topic familiarity in influencing learners' comprehension of AI-produced materials remains underexplored, leaving a gap in understanding how prior knowledge interacts with emerging content-generation technologies. Additionally, there is a pressing need for learner-centered research that investigates students' perceptions of AI-generated versus human-created materials to guide the design of effective and inclusive Open Educational Resources (OER). This study aims to address these gaps by empirically evaluating learner responses to both types of materials, examining the moderating effect of topic familiarity, and generating insights that can inform both theoretical discussions and practical applications in OER design and AI-assisted language pedagogy.

2.6. Key Research Questions

RQ1: To what extent does topic familiarity influence students' listening comprehension across AI-generated and human-created materials?

RQ2: How do college-level EFL students' listening comprehension performances differ when using AI-generated versus human-created listening materials?

RQ3: How do students perceive the difficulty level of AI-generated versus human-created listening materials?

The final question would be What implications do these findings have for the development and design of future OER?

3. Methodology

3.1. Participants

A total of 360 Chinese EFL students were recruited from a second-tier university located in central China to participate in this study. The participants were first-year students enrolled in second-semester College English Reading, Writing, and Translation courses at Henan University. Drawn from six separate classes of approximately 60 students each, the cohort represented a range of academic majors. At the time of the study, most students had not yet taken the College English Test

(CET) Band-4, a standardized assessment of English proficiency in China. Based on classroom performance and instructor assessments, most participants were estimated to be at the intermediate level (intermediate low or mid) according to the ACTFL (American Council on the Teaching of Foreign Languages) proficiency guidelines. Table 1 shows the number of students who took the listening comprehension exams and completed the survey for each topic. Some students didn't finish the exams, therefore they were not included in the listening comprehension exam data analysis.

Group	Noodle Exam	Noodle Survey	Panda Exam	Panda Survey	Hustle Culture Exam	Hustle Culture Survey	Cheese Rolling Festival Exam	Cheese Rolling Festival Survey
Class 1			56	59			57	57
Class 2			56	59			59	57
Class 3			52	54			55	57
Class 4	43	59			56	59		
Class 5	48	59			57	59		
Class 6	50	60			55	59		

Table 1. Number of Participants (out of 60 students each class)

3.2. Listening Materials

The topic classification (familiar vs. unfamiliar), as well as the human-created and AI-generated listening materials will be discussed in detail below. The study being described focused on how students responded to four different topics: *Panda, Noodle, Hustle Culture,* and *the Cheese Rolling Festival.* These topics were grouped into two categories based on how familiar they were to the students.

3.2.1 Familiar Topics – *Panda* and *Noodle*

These were considered culturally and contextually familiar to the students, who were native Chinese and attending Henan University.

- o *Panda*: The panda is native to China and holds significant cultural importance. It's often referred to as a national treasure. Because of its symbolic value and frequent presence in media, education, and national discourse, students would be expected to already have substantial background knowledge about pandas.

○ *Noodle*: Noodles have deep historical and cultural roots in China and are a staple in the Chinese diet. Henan province (where the university is located) is known for its noodle dishes, making them a part of everyday life for the students. Their familiarity with noodles would come not just from cultural knowledge, but also from lived experiences.

3.2.2. Unfamiliar Topics – *Hustle Culture* and *Cheese Rolling Festival*

These topics were considered less familiar or even unknown to the students.

○ *Hustle Culture*: This term refers to a mindset or lifestyle characterized by relentless work, ambition, and productivity, often popularized in Western (especially American) corporate or entrepreneurial contexts. It's not a concept commonly taught or discussed in Chinese educational materials or typical English curricula, so the students may not have encountered it before.

○ *Cheese Rolling Festival*: This is an unusual and specific event held in the UK, where participants race down a hill chasing a wheel of cheese. It's quirky and distinctly local, and like hustle culture, it hasn't been widely represented in Chinese educational content or cross-cultural discourse, making it likely unfamiliar to the students.

○ The classification of the topics was important for the researchers to analyze how students handle listening materials that are either within their cultural sphere or completely outside of it.

3.2.3. Audio Material Creation

Audio creation in this project involved four stages: (1) generating original recordings; (2) editing for EFL-appropriate length; (3) inputting the edited version into NotebookLM to create a podcast-style GenAI audio; and (4) refining the AI output to match the human version in length and focus.

Two audio tracks — one human-created and one AI-generated — were produced for each of the four topics, totaling four pairs. Human audios were in monologue or podcast form, created by individuals or groups. NotebookLM was used in April 2025, prior to Google's Gemini 2.5 Flash update in May. For *Panda*, the human audio was adapted from a 16-minute student presentation, edited to 6:26 minutes. NotebookLM generated a 30-minute podcast, later shortened to 20 minutes, and then refined into two shorter versions (9:24 and 10:20). Edits focused on retaining key content and removing repetition. The *Noodle* audio followed a similar process, starting from an 8-minute recording and edited to 5 minutes. The GenAI version ran 10:04 minutes, showing that AI output length did not directly match input length.

In contrast, *Hustle Culture* and *Cheese Rolling Festival* used original human podcasts with three speakers. After editing, *Hustle Culture* was reduced from nearly 53 minutes to 3:27, and *Cheese Rolling Festival* from about 30 minutes to 4:55, focusing on clarity and relevance. GenAI audios for these two topics were kept closer in length to their human counterparts and were shorter than the GenAI versions of *Panda* and *Noodle*. Research shows[1] that longer audio can hinder comprehension, especially for unfamiliar topics due to increased cognitive load. Based on this, longer audios were used for familiar topics, and shorter ones for unfamiliar topics in this study. It is worth pointing out that the overall lengths of the audio for familiar topics are longer than the unfamiliar topics, with the assumption that listeners have a higher tolerance for familiar topics than unfamiliar topics. The AI-generated audios covered all the key points from the human-created audios but included more supporting or engaging content between the speakers.

3.3. Procedure

Data collection took place over four non-consecutive days in May 2025. On Day 1, Classes 1-3 completed listening tasks on the topic *Panda*. Class 1 listened to the human-recorded audio first, followed by the GenAI version, answering the same set of comprehension questions after each. Class 2 followed the reverse order. Class 3 listened to a longer version of the GenAI audio twice, responding to the same questions each time; they did not hear the human version.

On Day 2, Classes 4-6 repeated the procedure with a familiar topic, *Noodle*. Class 4 listened to human audio first, then GenAI; Class 5 did the reverse; Class 6 listened to the same GenAI audio twice. On Day 3, Classes 4-6 returned to complete tasks on the unfamiliar topic *Hustle Culture*, following the same listening order as Day 2. On Day 4, Classes 1-3 completed

---

[1] Kim and Nam 2022.

the final task on another unfamiliar topic, *Cheese Rolling Festival*, using the same procedure as on Day 1, followed by a topic-related survey.

3.4. Instruments

Listening comprehension exams were given to the participants from each of the six classes. The students listened to two audio tracks in sequence and answered the same questions twice. The purpose was to see if the alternation of audio types (Class 1 and 2, Class 4 and 5) and repetition of an GenAI audio (Class 3 and 6) would affect their scores. There were five questions on each exam for each topic. They were multiple choice questions and true or false questions. For example:

1. Which of the following is true for the *Coopers Hill Cheese Rolling Festival*?
   a. Coopers Hill has a near vertical gradient with a one to two ratio and is 300 yards long.
   b. The cheese can reach speeds of up to 50 mph as it rolls down the hill.
   c. The prize for winning the *Coopers Hill Cheese Rolling* race is the cheese itself.
   d. The festival originated in France
2. Hustle culture is evolving to allow more personal priorities such as social life and health care.
   a. True
   b. False

The follow-up survey was designed to capture students' familiarity with the four topics and their perception of human-created content vs. GenAI audios. The first two questions on the survey addressed familiarity with the target topic Before and After listening to the audio track using a Likert scale: Q1: «On a scale of 1 to 10, how familiar are you with the topic of giant panda BEFORE today's task? Most familiar gets a 10 and least familiar gets a 1 and your rating is» and Q2 «On a scale of 1 to 10, how familiar are you with the topic of giant panda AFTER today's task? Most familiar gets a 10 and least familiar gets a 1 and your rating is». The third question (Q3) was to examine how much more knowledge about the topic they acquired after listening to the audio input. The fourth question (Q4) was an open-ended question on their perception of the difficulty level of these two types of audio inputs «In your opinion, which audio track human-created (track A) vs. GenAI-created (track B) is easier to understand and why?». The fifth (Q5) was to ask which was harder and why. The last question (Q6) was to ask the students to give opinions on seven statements using a Likert scale of A) Strongly Agree, B) Agree, C) Neutral, D) Disagree, and E) Strongly Disagree. The seven statements asked about students' perceptions of the two types of audio materials with respect to the speaker's pronunciation, intonation, etc.

3.5. Data Analysis

The students' listening comprehension results were analyzed quantitatively based on the test score in different listening sequence groups and across topics. The survey data was examined both quantitatively and qualitatively.

4. Results

4.1. Survey Results

To answer the first research question: to what extent does topic familiarity influence students' listening comprehension across AI-generated and human-created materials, we need to first find out students' perception of topic familiarity.

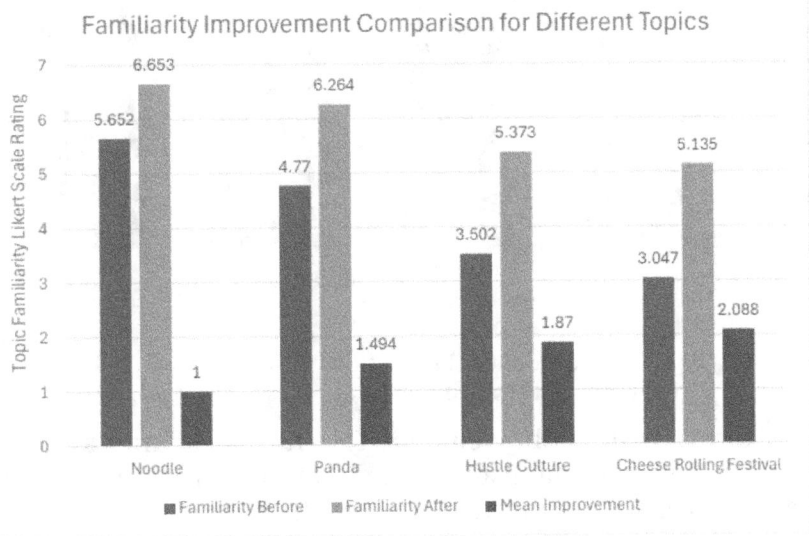

Figure 1. Students' Likert Scale Rating of the Familiarity of the Four Topics Before the Listening Exams

Figure 1. shows that 178 students from Class 4, Class 5 and Class 6 thought the topic of *Noodle* was most familiar to them with an average Likert scale of 5.652. The second most familiar topic was *Panda.* The 172 students from Class 1, Class 2, and Class 3 rated an average familiarity score of 4.77. *Hustle Culture* followed as the less familiar topic for the 177 students from Class 4, Class 5, and Class 6 with an average Likert scale score of 3.502. The least familiar topic for the 171 students from Class 1, 2 and 3 was *Cheese Rolling Festival.* The average score was 3.047.

After the listening activities, the familiarity ratings for all four topics were statistically significantly improved (Figure 1). The mean Likert score improvement for the four topics (*Noodle, Panda, Hustle Culture* and *Cheese Rolling Festival*) was 1 (17.69%), 1.494 (31.32%), 1.87 (53.39%) and 2.088 (68.52%).

However, when the types of audio materials and order of listening were considered, the level of familiarity improvement differed across three groups (Figure 2).

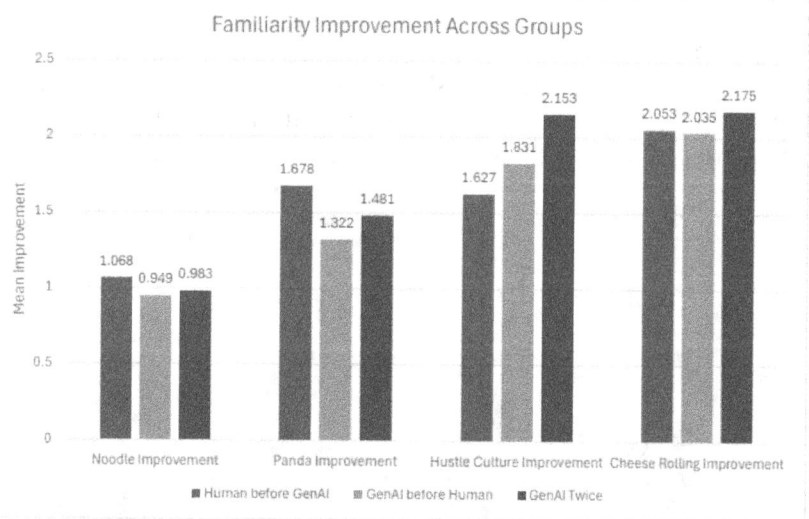

Figure 2. Students' Perception of Familiarity Improvement Across Groups

Survey results showed that for the most familiar topic, *Noodle*, students who listened to the human-created materials before the GenAI materials reported the highest improvement in familiarity among the three classes. The same pattern was observed for the next most familiar topic, *Panda.* Students who first listened to the human-created audio also showed the greatest improvement. For both topics, the students who listened only to the GenAI materials reported the second highest improvement. The group with the least familiarity improvement was the one that listened to GenAI materials before the human-created content.

For the less familiar topic of *Hustle Culture*, students who listened to the GenAI materials twice reported the highest improvement in familiarity improvement, followed by those who listened to the GenAI audio first and then the human-created audio. Students who listened to the human audio first and then the GenAI audio showed the lowest improvement. For the least familiar topic, *Cheese Rolling Festival*, the GenAI-only group again showed the greatest familiarity improvement. The group that listened to the human audio first ranked second, while those who listened to the GenAI audio first and then the human audio reported the least improvement. However, the difference in improvement between these last two groups was relatively small.

The third survey question (Q3) was designed to collect students' perception of how much they learned about new knowledge after listening to the audios. Class 1, 2 and 3 listened to the *Panda* and *Cheese Rolling Festival topics*, and Class 4, 5 and 6 listened to the *Noodle* and *Hustle Culture* topics. Most of the students agreed that listening to these audios helped them learn new things, ranging from 90% to 61% of students across the six classes.

Let's examine the two GenAI-only classes. While *Noodle* was the most familiar to students, 90% of Class 6 students who listened to the GenAI audio twice agreed or strongly agreed that they learned new information. Additionally, 81% of Class 6 students reported gaining new knowledge about the unfamiliar topic of *Hustle Culture*. These were the highest ratings among all 12 listening sessions. In contrast, fewer students from Class 3, who listened to the GenAI-generated content on *Panda* and the *Cheese Rolling Festival* twice, reported learning new things (61% and 68%, respectively). As for the Human-then-AI group (Class 4) and the AI-then-Human group (Class 5), they reported identical percentages of students who agreed or strongly agreed that they learned new knowledge about *Noodle* (71% vs. 71%) and *Hustle Culture* (63% vs. 63%). Similarly, the Human-then-AI group (Class 1) and the AI-then-Human group (Class 2) showed the same percentage of students (75% vs. 75%) who felt they learned something new about the familiar topic of *Panda*, and nearly identical percentages (72% vs.

75%) for the less familiar *Cheese Rolling Festival.* The order didn't seem to affect the students' perception in this regard.

When compared to the GenAI-only group (Class 3), the value of human-created content becomes more apparent, especially for culturally specific topics. A greater proportion of students in Class 1 (75%) and Class 2 (75%) reported learning new information about *Panda* than those in Class 3 (61%). A similar pattern emerged for *Cheese Rolling Festival,* with 72% of Class 1 and 75% of Class 2 students reporting new knowledge gained, compared to 68% in Class 3. Figures 3 and 4 illustrate two sides of the same coin: while GenAI excels at delivering knowledge on widely discussed global topics, human-created content remains more effective in conveying nuanced, culture-specific information.

The results of the next two survey questions Q4 and Q5 are presented by topic familiarity, from the most familiar *Noodle* to *Panda,* to the less familiar *Hustle Culture,* and finally to the least familiar topic, *Cheese Rolling Festival.* We will look at the three groups separately: first human then GenAI group, first GenAI then human group, and the GenAI only group. In the first group, more students thought the Human-created audios were easier for the familiar topics *Noodle* (58%), whereas for the unfamiliar topics *Hustle Culture* and *Cheese Rolling Festival,* more students (56%) thought the GenAI audio was easier. For the topic of *Panda,* there was no difference. When the order of the audio was reversed, a notable shift emerged in students' perceptions. Figure 6 shows that the students consistently reported that the human-created audio was easier to understand across all four topics. This was most evident with the most familiar topic, *Noodle,* where 69% of students found the human audio easier after first hearing the GenAI version. This perception also aligned with their test performance, as this group achieved the highest scores among the three *Noodle* exam groups. A similar trend was observed for *Hustle Culture* (56%), a broader cultural topic, and to a lesser degree for the more culturally specific topics *Panda* (53%) and *Cheese Rolling Festival* (51%).

However, when comparing the initial listening experience, students generally felt more comfortable starting with human-created audio than with GenAI. Across the four topics, 58%, 49%, 44%, and 42% of students, respectively, thought the human-created audio was easier, while only 29%, 47%, 44%, and

47% felt the GenAI audio was easier. This indicates a slight overall preference for human-created audio as the first exposure, especially for more familiar or accessible topics.

As for the GenAI-only group's perceptions of audio difficulty, the two most challenging topics were the culture-specific ones: *Panda* (93%) and *Cheese Rolling Festival* (89%). These findings align with responses to Survey Question 3, which asked about students' experiences learning new information from the audio, regardless of their prior familiarity with the topic. The results suggest that students' perception of audio difficulty was influenced more by the cultural specificity of the content than by how familiar they were with the topic itself. When digging into the reasons why they thought one audio was easier than the other, many students found that the human-created *Panda* audio was easier to understand because of its slower and clearer speech, which allowed them more time to process information and recognize individual words. They appreciated the logical and structural delivery of the content, which made it easier to follow the speakers' train of thought and identify key points. The absence of overlapping voices, filler words, or sudden interruptions also helped reduce distractions and allowed them to focus better. Additionally, students noted that they were more familiar with this type of single-speaker format, especially from past listening exercises in educational contexts. The steady, clear pronunciation, and consistent tone made human-created audio easier to follow and less mentally demanding to comprehend.

Interestingly, for those students who considered the GenAI podcast style audio was easier to understand, they listed the following reasons: 1) Speakers had clear pronunciation with a moderate pace and like other standardized delivery; 2) The dual-speaker format, often structured as a question-and-answer dialogue, made it easier to follow and identify key information. Apparently, the students were familiar with the format, which resembled the listening tasks they were used to in class or exams. The interaction between the two speakers was seen as more engaging and helpful for comprehension, allowing for natural pauses and clearer divisions in the content. Additionally, the GenAI audio was described as vivid, well-paced, and comfortable to listen to, with concise language and less ambiguity. These qualities made it easier for these students to stay focused and understand the overall meaning.

Negative opinions on the GenAI tracks included complaints about overlapping voices and interruptions among speakers, therefore the track required more attention to follow the speakers and the change of their tones. Interestingly, they found topic shifts between two speakers, filler words and interjections added difficulty in the podcast style audio track. One student indicated that listening to two people at once was mentally tiring, for instance, «Two people's conversation requires me to listen to both speakers, and it's hard for my mind to react». Structurally, it was noticed that the GenAI audio was more fragmented or less structured than the human-created monologue audio track. As for the human-created audio, the students who thought it was more difficult because the speech was too fast with few pauses, dense of formal vocabulary, unfamiliar pronunciation or speaking style and lack of interaction, which made it harder to stay focused or follow. Some found it boring or dry, reducing comprehension. For example, one student complained about the track being «too fast and the voice of the speaker was too different from what I was familiar with».

The final survey question examined nuanced distinctions between the human-created and GenAI-generated audio, focusing on aspects such as pronunciation, vocabulary usage, accuracy of information, speaking style (including speed and intonation), perceived information gaps, and students' preferences for future listening practice. Owing to space constraints, a detailed analysis of these findings will be presented in a separate publication.

4.2. Listening Comprehension Exam Results

Figure 3 presents the overall performance across the four topics, categorized by audio type. For both audio types, human-created or GenAI generated, students performed best on the *Noodle* topic, followed by *Cheese Rolling Festival*, then *Hustle Culture*, with *Panda* resulting in the lowest scores, except in the GenAI only group, the order was *Noodle, Hustle Culture, Cheese Rolling Festival* and then *Panda* for the test. Given that *Panda* is a culture-specific topic closely tied to Chinese identity, it may seem surprising that Chinese students performed the worst on this topic.

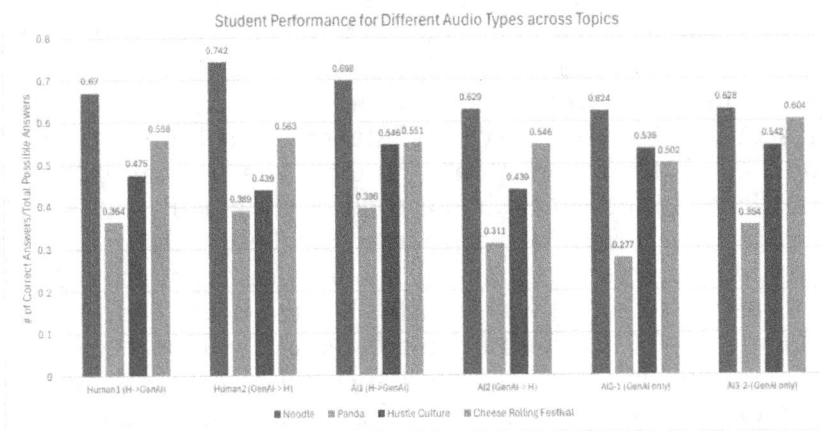

Figure 3. Overall Student Listening Comprehension Performance across Four Topics

Having examined the overall performance differences across topics, we now turn to a more detailed analysis of each individual topic, comparing results across the three exam groups: Class 4 was the first human then GenAI group, Class 5 was the first GenAI then human group, and Class 6 was the GenAI-only group. The proportion of correct answers out of the total possible was calculated. For instance, in the *Human2* test, Class 5 students correctly answered 178 out of a possible 240 questions (48 students × 5 questions), thus 0.742.

For the topic *Noodle,* among all groups, the highest average score (0.742) was achieved by students who listened to human-created audio after the GenAI audio (Class 5). The second-highest score (0.698) came from the group that listened to the GenAI audio after hearing the human-created version (Class 3). The group that listened only to GenAI audio (Class 6) performed the worst, with average scores of 0.624 and 0.628, even though more students (22%) from Class 6 rated *Noodle* as the easiest topic. Class 4 showed a slight improvement in performance, from an average score of 0.67 to 0.698. However, a paired t-test indicated that this improvement was not statistically significant ($t = 0.94$, $p \approx 0.35$). More obviously, Class 6's performance improvement was not statistically significant.

In contrast, Class 5 demonstrated a statistically significant improvement when moving from GenAI-generated to human-created audio. A paired t-test revealed a t-value of 3.698 with a p-value below 0.001, assuming a standard deviation of 1.0984 for the human-created audio scores. *Panda* was the next most familiar subject rated by students in survey question Q1. Among the lower-performing groups, Class 3 (GenAI-only) received the lowest score (0.277), while Class 1 achieved the highest score (0.396) after they first listened to the human-created audio and then the GenAI audio. The human-initiated knowledge input established a good starting point for Class 1 with a score of 0.364, which was the best starter. Class 2 showed a notable improvement, with scores rising from 0.311 to 0.389. This change was statistically significant, with $t(55) \approx 2.59$ and $p < 0.012$. The human-created audio appeared to significantly boost student performance, whereas the GenAI audio alone did not have the same effect. This result aligned with the previous GenAI first then human group (Class 5) for *Noodle*. For Class 1, the improvement from Human-created audio to GenAI audio was not statistically significant ($t(55) \approx 1.46$, $p \approx 0.15$), suggesting the slight increase may have been due to random variation rather than a reliable effect. Interestingly, although Class 3 initially performed the worst, their performance improved significantly after listening to the same GenAI audio a second time. This improvement was statistically significant, with a two-tailed p-value of approximately 0.0065 ($p < 0.01$).

When examining the performance scores for the topic *Hustle Culture*, Class 4 demonstrated a statistically significant improvement, $t(55) \approx 2.79$, $p<0.01$ (two-tailed $p \approx 0.007$). In contrast, Class 5 showed no measurable improvement, while Class 6 experienced only a minor, non-significant gain. These results suggest that for an unfamiliar topic, beginning with GenAI-generated audio may be particularly effective. For instance, Class 6 started with the highest initial score of 0.535, while Class 1, which began with human-created audio (0.475), showed a substantial improvement, reaching the highest score of 0.546 after listening to the GenAI version. However, Class 6 did not show significant improvement upon listening to the same audio a second time, indicating that repetition alone may not yield additional benefits.

Lastly, the performance scores were investigated for the topic *Cheese Rolling Festival*. It was noted that for this least familiar topic, Class 1 was the only group whose score became worse after the second audio was played, among all the 12 listening comprehension sequences across all the topics. Even though the drop from the sequence was not significant, from Human-created audio (0.558) to GenAI audio (0.551), it put AI audio's role in big question. Does AI work badly for unfamiliar topics or for the most culture specific topics? Previous results for *Hustle Culture* indicated the opposite for unfamiliar topics. GenAI audio worked quite well for it. It ruled out the first possibility. How about the second possibility? Was GenAI a low performing starter on more culture specific topics, such as *Panda?* It did! As a matter of fact, the group that listened to the GenAI panda audio for the first time performed the worst among all six exams that three *Panda* groups took. However, for Class 3 after the second time listening to the same *Cheese Rolling Festival* audio, its improvement was statistically significant. With a t(54)≈3.76 and p<0.001. This result echoed with the other culture specific topic *Panda's* GenAI only group.

5. Discussion

Survey results from Questions Q1 and Q2 revealed students were most familiar with the topic *Noodle*, followed by *Panda*, *Hustle Culture*, and least familiar with the *Cheese Rolling Festival*. Interestingly, the greatest improvements in familiarity occurred with the least familiar topics. This pattern suggests that the listening comprehension activities were effective in increasing students' topic familiarity, particularly for unfamiliar content.

A clear trend emerged regarding audio type, that is, human-created audio was more effective for topics students were already somewhat familiar with. GenAI-generated audio was more effective for initially unfamiliar topics. Survey Question Q3 asked students to reflect on how much new knowledge they gained. Responses indicated that using both human and AI audio enriched the learning experience by presenting content in different styles. However, the order of presentation (human first or GenAI first) did not significantly affect students' perception of learning gains. The performance of GenAI audio varied by topic type. It worked well with broad cultural topics such as *Noodle* and *Hustle Culture*, likely due to access to extensive general data. However, with culturally specific topics like *Panda* (Chinese culture) and *Cheese Rolling Festival* (British culture), GenAI-generated content tended to be less informative and offered fewer new insights.

This was especially evident with the *Panda* topic. Chinese students, already familiar with the subject, found the GenAI audio added little new information, which limited its impact. In contrast, the Cheese Rolling Festival — though also culturally specific — was less familiar to students overall, and comprehension improved with repeated exposure to GenAI content. Survey Questions Q4 and Q5 explored perceptions of audio difficulty. Students generally found that human audio was easier to understand when the topic was familiar. AI audio was more helpful for unfamiliar topics due to its standardized voice and clarity.

For unfamiliar topics, students who heard the human audio after the GenAI version showed better comprehension. In these cases, the GenAI audio acted as a primer, making the subsequent human version easier to process. However, when the human audio came first, the follow-up GenAI version did not improve performance significantly. These findings suggest that students' perception of difficulty was influenced more by the cultural specificity of the topic than by their initial familiarity with it. Listening comprehension scores did not fully align with familiarity rankings. Students scored highest on *Noodle*, followed by *Cheese Rolling Festival*, *Hustle Culture*, and lowest on *Panda*.

Other key findings include:

○ Repeating GenAI audio alone was not effective for the familiar topic *Noodle*.
○ Repeating GenAI audio was very effective for the challenging, specific topic *Panda*.
○ For *Hustle Culture*, starting with GenAI followed by human audio led to better performance than repeating GenAI audio alone.
○ For *Cheese Rolling Festival*, GenAI was less effective as a first exposure, but performance improved significantly with repeated listening.

Regarding the implications of these findings for creating OER listening materials, the first thing that comes to mind is to match audio type to topic familiarity. To use human-created audio for familiar topics is preferable, because human voices offer more natural variation in tone, pacing, and emotion, which can enhance engagement and reinforce existing knowledge. On the other hand, for unfamiliar or general cultural topics, the clarity and consistency of AI voices can reduce cognitive load of listeners, making them effective for initial exposure to new topics. OER content creators would also need to consider cultural specificality. Topics like *Panda* may require additional context or supplemental materials, especially using GenAI. AI-generated content may lack nuance or culturally accurate information, so pairing with curated human audio or visual aids can improve comprehension. It would be helpful to also include brief cultural annotations, vocabulary previews, or glossary sections for such topics to support learner understanding.

There are also two pedagogical implications for using OER content. Firstly, instructors may consider using sequencing strategically: start with GenAI audio to introduce new or unfamiliar topics and then follow with human audio for reinforcement and deeper comprehension. For familiar topics, either sequence may be used, but combining both types provides varied exposure, which can support different learning preferences. Secondly, instructors may leverage repetition, especially for difficult topics. Repeated listening to GenAI audio was shown to improve comprehension of difficult or unfamiliar topics. OER platforms can support this by 1) allowing easy replays of audio; 2) Offering slowed-down versions; 3) Including self-paced comprehension checks.

It is worth pointing out that data showed shorter audios (3-5 minutes) are more effective for unfamiliar or culturally dense topics to reduce cognitive overload, whereas longer audios can be used for familiar or general cultural topics where learners are more capable of sustaining attention and processing extended input. It would also help if OER platforms could provide students with options to choose audio types and adjust playback speed to support diverse learning needs.

6. Conclusion

In summary, the effectiveness of GenAI versus human audio depends on the topic type (general vs. culturally specific), students' prior familiarity, and the sequence of exposure. GenAI audio excels at introducing unfamiliar general topics, while human audio is more effective for reinforcing comprehension, especially when used after an initial AI-generated version. Culturally specific topics may require multiple exposures, regardless of the audio source, to support learner's understanding. The study highlights the importance of adaptive design in OER listening materials. By strategically combining human and AI-generated content, tailoring materials to topic familiarity and cultural complexity, and leveraging repetition and sequencing, educators can create OER that are not only accessible and scalable, but also pedagogically effective and culturally responsive.

One limitation of the present study lies in its sampling design: although all four topics were tested across six classes, each class was only exposed to two of the four topics. As a result, potential group differences were not controlled for or statistically accounted for in the data analysis. Additionally, the study did not include a condition in which students were exposed exclusively to human-created audio twice. This absence limited the ability to directly compare the effects of repeated exposure between GenAI-generated and human-created materials. Addressing these design limitations in future research would allow for a more robust comparison of audio types and their respective pedagogical impacts.

Bibliography

Bian, Xu, Xiaojun Cai and Dianmei Cai. 2019. «The Contributions of Listening and Reading Vocabularies to Listening Comprehension of Chinese EFL Students». *International Journal of Listening* 35.2: 110-22.

Chaikovska, Olha, Olha Stoliarenko, Nataliia Hlushkovetska and Iryna Semenyshena. 2024. «Enhancing Students' Listening Comprehension Skills through AI-based Podcast Activities: A Study in Self-Study Mode». *Environment Technology Resources* 2: 336-39.

EDUCAUSE. 2024. *Generative AI and the Future of OER: An EDUCAUSE Learning Lab.* Boulder: EDUCAUSE. ‹https://events.educause.edu›

Gilakjani, Abbas and Mohammad Ahmadi. 2011. «A Study of Factors Affecting EFL Learners' Comprehension and Strategies for Improvement». *Journal of Language Teaching and Research* 2: 977-88.

Hodgkinson-Williams, Cheryl. 2023. *Artificial Intelligence and Open Educational Resources in Africa: Opportunities and Challenges.* Johannesburg, OER Africa. ‹https://www.oerafrica.org›

Kim, Minkyung, Yunjung Nam and Scott A. Crossley. 2022. «Roles of Working Memory, Syllogistic Inferencing

Ability, and Linguistic Knowledge on Second Language Listening Comprehension for Passages of Different Lengths». *Language Testing* 39.4: 593-617.

Krashen, Stephen D. 1985. *The Input Hypothesis: Issues and Implications.* London, Longman. 92.

Lin, Yu-Ju and Hung-Chun Wang. 2018. «Using Enhanced OER Videos to Facilitate English L2 Learners' Multicultural Competence». *Computer & Education* 125.1: 74-85.

Mah, Boon Yih and Feifei Han. 2023. «Self-Assessed Online Tests as an OER: the Senior English Educators' Perspectives Based on UTAUT Model». *International Journal of Information and Education Technology* 13.7: 1135-41.

Schmidt-Rinehart, Barbara C. 1994. «The Effects of Topic Familiarity on Second Language Listening Comprehension». *The Modern Language Journal* 78.2: 179-89.

University of Virginia Library. 2024. *OER + AI Tools Guide.* Charlottesville, University of Virginia. ‹https://guides.library.virginia.edu/oer-ai›

Voice and Authorship in the Era of AI: Implications for Second Language Writing Teachers

Gergana Vitanova
and Aimee Schoonmaker

Authorship is a complex topic. «Who can be an author?» is a question that could be answered in multiple ways. Some[1] adopt Bakhtin's (1984) dialogical framework, in which the writer or the speaking subject co-creates meaning through interactions with others. Moreover, in Bakhtin's democratic approach, anyone can author their own words and worldviews, thus giving unique agency to humans. In addressing authorship, Bakhtin certainly did not envision human interaction with artificial intelligence (AI), which challenges our understanding of creativity and ethics. At least one CNN article[2] warns that AI is producing and falsely presenting books as written by actual human writers. How do we define authorship in the context of this fast-developing technology?

If this question is thorny when we discuss professional writers who are experts in the use of language, the implications for second language (L2) writers and their teachers become even more numerous. Voice and authenticity are essential in L2 writing.[3] The processes of writing in a first and second language are also different. Writing in a second language presents not only linguistic (e.g., vocabulary, syntactic, etc.) challenges but also difficulties related to the lack of awareness of genres, culture, and even content. It has been well established[4] that writing is a process, one that entails planning, drafting, multiple revisions, and peer feedback. Yet why would L2 writers go through these multiple stages of drafting if they

---

[1] Vitanova 2010.
[2] Duffy 2023.
[3] Hirvela and Belcher 2001.
[4] Grabe 2001.

could generate an essay quickly and effortlessly? More importantly, how will writers, who do not have the linguistic and cultural tools to evaluate texts produced by ChatGPT, acquire the skills they need to compose academic texts independently?

While some research on the use of AI in language teaching is beginning to emerge,[5] L2 writing teachers' perceptions of this new tool have not been explored. In this paper, we discuss the notion of voice in L2 writing and the challenges that tools such as ChatGPT present to language learners and teachers. We specifically focus on the lived experiences of a seasoned L2 teacher in an English for Academic Purposes (EAP) program. Her critical reflections and observations of practices by students will be employed to illustrate the opportunities and potential pitfalls that AI creates for L2 writing classrooms. In the end of the paper, we share our recommendations for second language teachers.

Authorship, dialogue, and voice

The notion of dialogue has been one of the most significant concepts in Russian philosopher Bakhtin's work. In Bakhtin's words, «The dialogic orientation of discourse is a phenomenon that is, of course, a property of any discourse».[6] Bakhtin, who wrote about the genre of the novel, did not, however, limit the notion of dialogue to relationships within the text or within the discourse of two human beings interacting with each other. Instead, dialogue is key in human thought and existence. Dialogue, to Bakhtin, permeates all human existence and discourse:

> Dialogic relationships are a much broader phenomenon than mere rejoinders in a dialogue laid out compositionally in the text; they are an almost universal phenomenon, permeating all human speech and all relationships and manifestations of human life in general, everything that has meaning and significance.[7]

Knowledge itself is a shared human experience in this framework. For example, Hirschkop argues that dialogue refers to the exchange of ideas and humans' world views. Bakh-

---

[5] Pack and Maloney 2024.
[6] Bakhtin 1981, 279.
[7] Bakhtin 1984, 40.

tin's complex framework of dialogism has been applied to second language learning in various ways.[8] Vitanova, for instance, employed it to investigate how agency develops in several language learners and immigrants in the United States. In that, she drew both on the notions of dialogue in discourse and on creative responses (or answerability) that underlie human acts. Anyone who speaks can be considered an author as he postulates in his book *Toward the Philosophy of the Act*.

In L1 (first language) academic writing, Lillis, drawing on Bakhtin's dialogism, argued for moving from monological to dialogical approaches in writing pedagogy. In second language writing, in which the notion of voice has become increasingly important, Prior has cogently articulated the need to include Bakhtin's sociohistoric theory in the conceptualizations of discourse and voice. Building on Bakhtin's framework,[9] in his now classic article, Prior claims that discourse is always social and historical and provides the following understanding of the somewhat elusive notion of voice: «Voice is a typification linked to social identities; voice as the reenvoicing of others' words (oral and written) through processes of repetition and presupposition; and finally, voice as it is linked to situated productions of persons and social formations».[10]

Tardy, tracing the development of voice in L2 writing, notes its many dimensions, for example, voice as an aspect of identity, voice as a unique individual expression (a perspective that teachers often employ), and voice as interacting with genres and discourses. Thus, voice is strongly related to writers' complex social and textual identities. Voice is also dialogical and relational. When writers create or author a text — in L1 or L2 — they are drawing on the voices of others they have encountered in the past from various discourses and genres. In this sense, the writer's voice is «an amalgamated dialogical effect».[11] Embracing voice as a dialogical construction, Canagarajah[12] suggest that L2 writing teachers will have to address the following questions as they design their courses:

1.   What are the components amalgamated in voice?

---

[8] Hall, Vitanova and Marchenkova 2005; Vitanova, 2010.
[9] Bakthin 1981; 1984.
[10] Prior 2001, 55.
[11] Canagarajah 2015, 123.
[12] Canagarajah 2015, 123.

2. What is the nature of the negotiations that characterize dialogical voice?
3. How do interlocutors (i.e., teachers, peers) mitigate their appropriation of writers' voices in the achievement of 'effect'?[13]

Canagarajah's questions are challenging to address even without the presence of AI. Current L2 writing approaches maintain that teachers should encourage writers' authentic voices. Yet, how do their voices remain unique, or, how do they author themselves in an era when AI is evolving rapidly? What would the components of their amalgamated voice be, given that the nature of AI itself seems to be one ever increasing the amalgamation of voices and discourses? While research on the topic is still very limited, the following section briefly looks at some of the emerging thoughts on AI in L2 classrooms.

AI in language learning and L2 writing

The topic of AI has emerged as significant in most educational forums and conversations. The ways in which it has infiltrated daily educational practices affected how we look at pedagogically sound endeavors and potentially changed students' views on what learning is. For instance, at the recent TESOL (Teaching English to Speakers of Other Languages) 2025 international convention, as well as at the American Association for Applied Linguistics conference, AI was the focus of many presentations and discussions, and although there were no definite answers to all that it entails, it generated a vibrant dialogue.

As students all over the world have been intensely exposed to AI's myriad possibilities, educators have promptly immersed themselves into the current discussion to understand and boost their own AI literacy, enhance engineering proficiency, and more than ever, ensure that robust critical thinking skills will keep the transformative aspects of this new reality in check.[14] The notion of approaching AI as an ally (not a threat) might be the inevitable and wiser route to overcome the fear and anxiety many feel about its unquestionable and irreversible power. Yet, many questions arise about the ethical elements of AI, the need to constantly evaluate its use, and how

---

[13] Canagarajah 2015, 123.
[14] Walter 2024.

to draw an ethics-oriented line between the acceptance of this new reality while devising strategies to foster pedagogically sound practices and critical thinking opportunities for students. For instance, ChatGPT has been one of the most common AI tools used in education. While scholars recognize some of its potential in writing, others[15] worry that it may impede students' critical thinking skills or creativity.[16]

Wibooalysarin[17] investigated how AI, through ChatGPT-generated prompts, could enhance collaborative writing and written corrective feedback by focusing on exchange students learning Thai. Among the benefits of employing AI the authors mention that AI can be used as an initial screening tool that allows instructors to deal with more important pedagogical matters. Additionally, AI could potentially provide feedback that is tailored to students' linguistic needs. They also include a few concerns, however, which could include the over-reliance on AI-generated feedback and, thus, the lack of more nuanced feedback provided by instructors who know their writers' needs best.

In one particularly relevant to our topic study, Derakhshan and Ghiasvand[18] investigated the perceptions of 30 EFL (English as a Foreign Language) teachers in Tehran. Specifically, they focused on these teachers' perceptions of the impact of ChatGPT on assessment and L2 teaching. Some of the benefits the participants associated with ChatGPT were the potential for learner autonomy, the tool's being constantly available to learners, as well as its potential to provide immediate feedback. Among the major pitfalls the teachers reported were the potential compromise of academic integrity, as well as ChatGPT's promoting superficial understanding of concepts and shallow literacy skills overall.

Derakhshan and Ghiasvand's findings are not going to be a complete surprise to L2 writing teachers as they were familiar with some of these challenges even before the emergence of ChatGPT. Both of the authors have taught L2 writing, and, thus, have witnessed how the use of technology could be beneficial to L2 writers in many ways. For example, it can help with the research of a particular topic. On the other hand, we have also

---

[15] Susnjak and MacIntosh 2024.
[16] Buriak 2023.
[17] Wibooalysarin 2024.
[18] Derakhshan and Ghiasvand 2024.

observed in our own courses how tempting it could be for students to copy readily available passages in some of the materials available online. With ChatGPT, integrity is not the only aspect that teachers could find challenging. They may also have to navigate the difficult issues of authorship of the papers produced by their students.

It is fair to say that, amidst the many questions and discussions AI has generated in recent months, the wisest route to it is one of a solid exchange of views and gathering of «a holistic and trustworthy picture»[19] of its use in education. As we look at different angles of the AI spectrum in the higher education realm, the many inquiries it has been triggering are bound to remain unanswered for now, and it seems clear that the fast pace in which AI has been moving points to a more established temporary stage of uncertainty. Asking, listening, debating and being open to possible adaptations in the classroom are actions that many feel under pressure to take, and some practices that would seem unthinkable mere months ago are now being considered, as instructors struggle to find ways to incorporate AI tools into their practice. Such a quest to use AI tools could suggest an urgency from teachers to adopt them, partly as a genuine wish to understand and use AI in pedagogically robust ways, but also partly due to an unspoken need to prove that they are not reacting negatively to the newest technological trend and that they are able to handle it. Yet, the navigation into these waters is not without turbulence and at every attempt to master AI-oriented activities, new layers of inquiries arise.

One teacher's experience with AI in L2 academic writing

One of us, Schoonmaker, teaches EAP (English for Academic Purposes) writing courses for a population of international students in a large urban research university. In November 2022, as these EAP students worked on the detailed revisions of their final essays, and after a term that revolved around several proofreading stages of these writings, the talks about AI and ChatGPT started doing the rounds in academia. By spring 2023, the initially distant notion of a very effective AI tool that could write (supposedly) better than human beings started materializing through concrete cases of EAP students

---

[19] Walter 2024, 2.

submitting writing work that was clearly, and astonishingly, superior to what they were able to do mere weeks earlier. What was happening?

Taking a brief look at the background for this particular EAP course, students were mainly focused on writing response essays, where they would read a text about writing and react to it by addressing specific prompts, and under a Writing About Writing (WAW) pedagogy.[20] WAW proposes that writing experts write about their craft, the act of writing, and its myriad intricate elements. Such an approach gives students both the exposure to a higher level of writing from writing scholars, as well as the freedom to write about how they felt towards the text and the authors' ideas and perspectives, at the same time making clear connections and applications to aspects such as their own writing situations, audience, and genre, among others. Under that prerogative, what the instructors in the EAP program had been doing for several years at that point was a gradual and detailed construction of short essays that were created from the introduction to the body paragraphs and finally the conclusion, with consistent feedback for each one of those stages. The idea was to give students the time and ample support to build their understanding of the text, how to react to it, and how to organize their own ideas in a text, at the same time looking at concepts of coherence, cohesion and objectivity while answering the proposed prompts.

The systematic corrective feedback students receive addresses several aspects of their writing, such as grammar, vocabulary, essay organization, coherence, cohesion, clarity, and the text ideas in general, but at its core the feedback aims at encouraging students to reflect upon their initial writing choices, analyze them with fresh eyes, and gradually build on what was done prior to receiving said feedback. As opposed to an ethnocentric view that determines a definitive right or wrong way to write, students are invited to perceive themselves as authors, and as such, agents of their own learning and writing journeys. The feedback process here espouses the notion of voice directly aligned with Prior's view of productions of persons and social formation.[21]

---

[20] Wardle and Downs 2020.
[21] Prior 2001.

The dialogue between the student and the instructor does not imply a superiority on the part of the instructor, but a collaboration and a social and academic interaction on both parts, where comments, suggestions, and positive reinforcement of what they are doing as writers will create a rich and empowering environment for them to grow and become better writers. Moreover, students are encouraged to understand how other authors build their argument as well as make their own rhetorical choices with informed opinions and critical thinking. In other words, in mastering the conventions of academic writing, students are also encouraged to develop their own authentic voices. Different types of dialogue, including dialogue between students as active readers and texts, dialogue with the instructor, and dialogue with other fellow writers through peer work, were fostered as well.

Schoonmaker recalls that, in spring 2023, AI was suddenly brought into the EAP environment, and instructors started noticing some unusually well written texts submitted by students whose original diagnostic tests were significantly lower in quality and depth, as well as lexical choices and overall essay organization. While plagiarism is not a new issue in classrooms, the EAP instructors were faced with a new challenge. At the time, at the very beginning of the AI phenomenon in academics worldwide, the initial biggest challenge was how to deal, in practical terms, with the cases of students using AI, from a plagiarism perspective. The software the institution used (Turnitin) was also dealing with the aftermath of AI by adding a new tool within their detection software that would supposedly point to AI issues. Instructors very quickly realized that such a tool was neither efficient nor accurate. Institutions may not know how to support writing teachers in this case, either. When consulting with the office that deals with plagiarism at the university, instructors were advised to avoid judging and penalizing students, and, instead, to find a one-on-one dialogical approach that would allow students to feel comfortable to admit they used it, and by doing so we, as faculty, would take the necessary measures, often using an AI plagiarism situation as a learning opportunity.

One specific situation that happened during this phase involved a student who was doing poorly in the course overall, and whose writing assignment was flagged by the Turnitin software as fully AI. He had done the first portion of the essay, the introduction paragraph, in the classroom and by hand. However, as he submitted the next paragraphs, which had been done at home on a computer, the work was flagged as completely done by AI. Schoonmaker knew that was not an accurate report for the introduction paragraph, but she also knew that the remaining paragraphs were way above his writing capabilities at the time. The grammar was perfect and the lexical choices were very sophisticated, and this particular student's writing was not at that level so early in the course. He was invited for a conference in Schoonmaker's office, and after some reassurance that her goal was to educate him on the consequences of submitting work that was not his, he admitted that he had used AI to finish his work, advised by a fellow classmate.

This case had a two-fold outcome that was mostly positive. It taught us that the detection software that was used was clearly flawed, as it included his first paragraph in the report as also being generated by AI, and we knew for sure that was not the case. It was also a positive learning lesson for the student. Although he did not receive any points for the AI portions, he was allowed to do it again for partial points, and our conversation reinforced notions of academic honesty and the need to submit work that is authentic, produced by him, and at his level of proficiency.

After the initial phase described above, L2 writing instructors started looking at the issue of AI from a different perspective, moving from a panic/avoidance attitude to one that seeks innovative ways that would 'incorporate' AI to the writing realm while maintaining pedagogically-sound opportunities for critical and ethical thinking. It was fundamental to educate students not only to improve their digital literacy and prompt engineering, but to also strengthen their understanding that AI should not do their work for them, but support their development as writers. By doing that, we instilled in students the need to develop their own unique voices as writers, to be agents of their own learning and master AI available tools without becoming dependent on them.

Workshop

In the section that follows, we illustrate how Schoonmaker addressed the use of AI in an L2 writing context pedagogically. As part of the effort to strengthen students' confidence about their own writing capabilities and build on their understanding of AI in the educational realm, the EAP team of instructors designed a workshop that would give students the opportunity to talk openly about AI, learn about its possibilities in a writing course, understand what AI prompt engineering is, learn how to analyze AI output with critical eyes, and exercise their critical thinking skills about that topic.

The workshop was prepared with three main objectives in mind:1) evaluate AI output by analyzing AI-generated text samples for accuracy, relevance, and potential biases; 2) understand and apply prompt engineering: a guide to understand the several types of prompt and how to refine AI output for writing tasks; and 3) encourage AI use that is responsible and ethical, avoiding plagiarism in the process. The workshop was scheduled to be conducted ideally over two class periods: Day 1: AI and critical thinking and Day 2: prompt engineering and output analysis. Both a pre-workshop survey as well as a post-workshop reflection was conducted as part of the process.

Pre-workshop

Prior to Day 1, students were sent a link to take a Pre AI Questionnaire at home. The survey asked students the following questions: What is your major? What is your first language? How would you rate your academic writing in English? Have you used Artificial Intelligence (AI) apps (also called large language models – LLMs) before? (e.g., ChatGPT, Claude.AI, CoPilot, Bard/Gemini). If yes, which AI Chatbots (LLMs) have you used? (You can select more than one option.) What do you use AI for? What do you think about university students using AI for academic writing? How comfortable are you with using AI tools (like ChatGPT) for writing assistance? The idea for the questionnaire was to gauge levels of familiarity students had with the AI tools so instructors could pair students in ways that could provide a meaningful exchange of experiences among them.

Workshop Part I: Critical Thinking Discussion

On the first workshop day, students were paired according to the answers from the questionnaire, and started with a Critical Thinking activity where they had to discuss the following questions: What are your initial thoughts on AI? What is it, and

how does it work? In what ways do you think AI could be helpful or harmful in academic writing? Would you consider using
AI to help you with your writing? Why or why not?

The answers were discussed then by the whole class and
the instructors commented on the differences in perspectives
and opinions. It was interesting to notice the very different approaches different students had on AI, as predicted. After that,
several elements of AI such as fact checking and hallucinations, bias, grammar and vocabulary levels, MLA formatting
issues, among others were defined and explained.

Workshop Part II: Analyzing AI-Generated Content

On the second workshop day students received instruction
on the types of prompts they can generate with AI: zero shot,
few-shot, and role play. They were then given two samples of
writing, one AI-generated, and one produced by an EAP student about the same text and the same prompt. They had to
analyze these two samples and define which one was AI and
which one was human-generated. They were encouraged to
look at elements such as fact checking, language level match in
terms of both vocabulary and grammar, clarity, coherence and
cohesion, and finally MLA citations accuracy. They received a
physical copy of the samples but could also access the samples
virtually. The analyses were done in pairs to generate a richer
exchange of ideas and perspectives, and to allow them to benefit from each other's experience or lack thereof regarding AI
technology. As previously mentioned, some students were significantly more advanced than others in their AI use.

The discussion that followed the analyses pair work was
engaging and enthusiastic: students were excited to point out
to instances in which the AI text was less than ideal, extrapolating the original text's content to produce some improbable
(and very wordy, somewhat artificial) sentences, using vocabulary that was excessively formal or clearly above their level,
and either lacking accuracy in the citations, or simply not using citations at all.

Post-workshop reflection

After the class discussion on the samples, students answered the following questions on a post-workshop survey:
What surprised you the most while analyzing the AI-generated
essays? How did this activity change your perception of AI and

writing? Why is it important to be a critical thinker when using AI? The answers were in several instances critical of AI as a perceived ideal, perfect tool that will give students the best product for their academic pursuits. In some cases, students demonstrated surprise that AI can be fallible and that trusting AI as a sole source of output can be a debatable decision, but also pointed out that AI can be a helpful tool if used with a grain of academic salt.

Conclusion

Dialogue, voice, and authorship are central in L2 writing. However, yet another voice has been added to the already complex environment of EAP classrooms. AI has become ubiquitous and seems to be here to stay. Theories and research are still emerging to suggest how it would affect L2 writers and teachers. For ESL learners, writing skills are critical for academic success, and the emergence of generative AI tools is opening opportunities for educators to seek solutions that are both ethically and pedagogically sound,[22] and that can contribute to the improvement of students' writing skills in myriad ways. Yet, it has also brought challenges. Aside from plagiarism, one concern is whether writers would acquire the skills that would allow them to evaluate texts produced by the relatively new technology. How would L2 writers, who are still developing their own second-language voices and are practicing new linguistic and rhetorical structures, dialogue with the rapidly evolving tool? This is perhaps the most important challenge L2 teachers face.

These uncertainties speak to the need to educate both L2 writers and instructors. In practical terms, L2 teachers have to create opportunities for students to think critically about AI, test its capabilities, detect its flaws, and understand its limitations. One specific approach is to conduct workshops or even mini-lectures in class to educate students on AI, demonstrate to them that AI is not always appropriate for all tasks and exemplify ways in which it can be used in a pedagogically sound and ethical fashion, supporting them in becoming better writers, rather than doing their writing for them. In other words, AI does not have to become an unquestioned, shadow author of their voices and texts. Instead, it can be a supplementary

---

[22] Pack and Maloney 2024.

tool that allows them to engage in dialogue with other voices and worldviews as well as to critique them.

In this new perspective, AI becomes one more interlocutor with whom writers can interact and co-create meaning, similar to what Bakhtin proposed in his dialogical framework (1984). Even more, depending on how workshops on prompt engineering and AI literacy are conducted, they can open up a wide variety of activities, such as AI prompt output analyses and grammar error detection, which can foster students' critical thinking abilities. It is a view that embraces the inevitability of AI incorporation into the academic realm without giving up the notion of community-built text that enhances human interaction (1984). The new element is that AI has become part of this polyphonic effort.

Bibliography

Bakhtin, Mikhail. 1981. «Discourse in the Novel». In *The Dialogic Imagination: Four Essays*. Michael Holquist, ed. Austin, University of Texas. 259-422.

———. 1984. *Problems of Dostoevsky's Poetics*. Minneapolis, University of Minnesota Press.

———. 1993. *Toward a Philosophy of the Act*. Austin, University of Texas Press.

Buriak, Jillian *et al.* 2023. «Best Practices for Using AI when Writing Scientific Manuscripts: Caution, Care, and Consideration: Creative Science Depends on It». *ACS Nano* 17: 4091-93.

Canagarajah, Suresh. 2015. «"Blessed in My Own Way": Pedagogical Affordances for Dialogical Voice Construction in Multilingual Student Writing». *Journal of Second Language Writing* 27: 122-37.

Derakhshan, Ali and Farhad Ghiasvand. 2024. «Is ChatGPT an Evil or an Angel for Second Language Education and Research? A Phenomenographic Study of Research-Active EFL Teachers' Perceptions». *International Journal of Applied Linguistics*: 1-19.

Duffy, Clare. 2023. «An Author Says AI is "Writing" Unauthorized Books Being Sold Under Her Name on Amazon». *CNN*. 10 August 2023. ‹https://www.cnn.com/2023/08/10/tech/ai-generated-books-amazon/index.html›

Grabe, William. 2001. «Notes Toward the Theory of Second Language Writing». In *On Second Language Writing*.

Tony Sylva and Paul Matsuda, eds. New York, Routledge. 39-57.

Hirschkop, Ken. 1998. «Is Dialogism for Real?». In *The Contexts of Bakhtin: Philosophy, Authorship, Aesthetics.* David Shephard, ed. New York, Routledge. 183-95.

Hirvela, Alan and Diane Belcher. 2000. «Coming Back to Voice: The Multiple Voices and Identities of Mature Multilingual Writers». *Journal of Second Language Writing* 10.1-2: 83-106.

Hall, Joan Kelly, Gergana Vitanova and Ludmila Marchinkova, eds. 2005. *Dialogues with Bakhtin: New Perspectives on Second and Foreign Language Learning.* Mahwah, Lawrence Erlbaum.

Lillis, Theresa. 2003. «Student Writing as 'Academic Literacies': Drawing on Bakhtin to Move from Critique to Design». *Language and Education* 17.3: 192-207.

Pack, Austin and Jeffrey Maloney. 2024. «Using Artificial Intelligence in TESOL: Some Ethical and Pedagogical Considerations». *TESOL Quarterly* 58.2: 1007-18.

Prior, Paul. 2001. «Voices in Text, Mind, and Society: Sociohistoric Accounts of Discourse Acquisition and Use». *Journal of Second Language Writing* 10: 55-81

Susnjak, Teo and Timothy McIntosh. 2024. «ChatGPT: The End of Online Exam Integrity?». *Education Sciences* 14.656: 1-20. ‹https://www.mdpi.com/2227-7102/14/6/656›

Tardy, Christine. 2016. «Voice and Identity». In *Handbook of Second and Foreign Language Writing.* Manchon, Rosa and Paul Kei Matsuda, eds. Boston, de Gruyter. 349-63.

Vitanova, Gergana. 2010. *Authoring the Dialogical Self: Gender, Agency and Language Practices.* Amsterdam, John Benjamins.

Walter, Yoshija. 2024. «Embracing the Future of Artificial Intelligence in the Classroom: The Relevance of AI Literacy, Prompt Engineering, and Critical Thinking in Modern Education». *International Journal of Educational Technology in Higher Education* 21.15: 1-29. ‹https://educationaltechnologyjournal.springeropen.com/articles/10.1186/s41239-024-00448-3›

Wardle, Elizabeth and Doug Downs. 2023. *Writing about Writing.* New York, Macmillan Learning.

Wibooalyasarin, Watchaparoi, Kanokpan Wiboolyasarin, Kanpabhat Suwanwihok, Nattawut Jinowat and Renu

Muenjanchoey. 2024. «Synergizing Collaborative Writing and AI Feedback: An Investigation into Enhancing L2 Writing Proficiency in Wiki-based Environments». *Computers and Education: Artificial Intelligence* 6: 1-10. ‹https://www.sciencedirect.com/science/article/pii/S2666920X24000298?via%3Dihub›

Contributors

Susana L. M. Antunes is an Associate Professor in the Department of Spanish and Portuguese at the University of Wisconsin–Milwaukee, where she teaches Portuguese language as well as Lusophone literature and culture. Her research explores contemporary poetry, travel writing, island literature in Portuguese-speaking contexts (including Brazil and Africa), geopoetics, and ecocriticism. She co-edited *Trinta e Muitos Anos de Devoção: Estudos Sobre Jorge de Sena em Honra de Mécia de Sena* (Ver Açor, 2016) and *Rememorando Daniel de Sá: Escritor dos Açores e do Mundo* (Ver Açor, 2016). She is the author of *De Errâncias e Viagens Poéticas em Jorge de Sena e Cecília Meireles* (Afrontamento, 2020), and editor of *Ilhas de vozes em reencontros compartilhados* (Quod Manet, 2021).

Martha Brenckle received her BA and MA from Southern Connecticut State University in New Haven, and in 1993, earned her PhD from Lehigh University in Bethlehem, PA. She is a Professor at the University of Central Florida in the Writing and Rhetoric Department and affiliated faculty in Women and Gender Studies and Text and Technologies, a transdisciplinary PhD Program in the College of Arts and Humanities. Dr. Brenckle teaches undergraduate and graduate courses in Rhetorical Theory. Her research involves First-year Writing, Feminism, and Queer Theory. She has published in *College Composition and Communication, Composition Studies,* and the *Journal of Basic Writing*. Dr. Brenckle also publishes poetry and fiction and has been nominated for a Pushcart Prize, a Triangle Award and a Lambda Award. Brenckle's poetry manuscripts include *Hard Letters and Folded Wings* (Finishing Line Press, 2019) and *Inside a snow globe* (Moonstone Arts Press, 2025).

Patricia Farless is a Senior Instructor at the University of Central Florida where she has taught courses in American History, US Women's History, and US Legal History. Her research interests include US Women's Legal History, Constitutional History, 19th-Century History, Women's History, the Civil War, and Reconstruction. Ms. Farless earned a BA in History and Political Science and an MA in History. She is the General Education Coordinator for the Department of History. Additionally, she coordinates and advises the Prelaw in the Humanities Minor at the University of Central Florida.

Flavia Azeredo-Cerqueira is an Associate Teaching Professor of Portuguese and Director of the Portuguese Language Program in the Department of Modern Languages and Literatures at Johns Hopkins University. She earned her Ph.D. in Applied Linguistics, with a concentration in Second Language Acquisition (SLA), from the Federal University of Minas Gerais (UFMG) in Brazil. Dr. Azeredo-Cerqueira's scholarly work encompasses a broad spectrum of SLA-related topics, including language identity, learner motivation, corrective feedback, phonetics and phonology, teacher training, telecollaboration, and the impact of study abroad experiences on language development and intercultural competence. Her current research focuses on the intersection of language learning and emerging technologies, with a particular emphasis on the potential of Artificial Intelligence (AI) to enhance second language acquisition.

Eduardo Viana da Silva is Teaching Professor and Coordinator of the Portuguese Program at the University of Washington, Seattle, United States. He received his Ph.D. in Luso-Brazilian Literature with an emphasis in Applied Linguistics from the University of California, Santa Barbara (UCSB). He also holds an interdisciplinary Certificate in College and University Teaching from UCSB. His research focuses on applied linguistics, environmental studies, virtual exchanges, and Brazilian literature and art. He has recently conducted study abroad programs to the Pantanal region of Brazil focusing on art, environment, and activism. He is also interested in how the development of Artificial Intelligence (AI) will enhance the teaching of languages and the possible limitations and challenges of AI on the teaching of Portuguese.

Karina Lissette Cespedes is an Associate Professor in the Department of Philosophy, Humanities and Cultural Studies at the University of Central Florida. Her research interests include relational reflexivity and pedagogy, Cuban Studies, tourism, food security, and art. Most recently, Karina has published on the afterlife of Cuban slavery while writing her book on Cuban food insecurity, *Hunger in Havana*, under contract with Vanderbilt University Press. She has authored and co-authored numerous articles and book chapters.

James R. Paradiso is a senior instructional designer at the University of Central Florida. His main areas of research and professional specialization involve investigating how principles from the learning sciences and data analytics can be applied to improve educational outcomes at the postsecondary level.

Chloë Rae Edmonson is Assistant Professor of Theatre at Trinity University in San Antonio, Texas. She is the author of America Under the Influence: Drinking, Culture, and Immersive Theatre (Routledge 2023) and her work has appeared in Performance Research, The Drama Review, and Performance Matters. In addition to her scholarship, Edmonson is a dramaturg and director, with interests in immersive theatre, artificial intelligence, and the historical avant garde.

Julia Listengarten is Professor of Theatre and Artistic Director in the School of Performing Arts at the University of Central Florida. She is an artist-scholar whose research explores avant-garde and contemporary theatre, interdisciplinary artistic practice, and socially engaged performance that critically engages with issues of precarity and marginalization. As a director and dramaturg, she has contributed to the development of new works through new play workshops and festivals. Among her recent publications are Visual and Performing Arts Collaborations in Higher Education: Transdisciplinary Practices (Palgrave 2023), Performing Arousal: Precarious Bodies and Frames of Representation (Methuen Drama/ Bloomsbury 2022), and The Cambridge Companion to American Drama since 1945 (Cambridge UP 2021). She co-edited the 8-volume book series Decades of Modern American Playwriting: 1930-2009 (Methuen Drama/ Bloomsbury 2018) and is currently co-editing (with Alissa Clarke) the book series Reflections on Contemporary Performance Process for Bloomsbury Publishing.

Alyssa Barrack is a PhD student in the Texts and Technology program at the University of Central Florida. Identifying as a «jack of all trades», Alyssa has professional experience in event management, production dramaturgy, scenic artistry, lighting design, professional writing, and adjunct teaching. With a deep commitment to inter- and transdisciplinary research, Alyssa's work explores audiences in theatre, film, and digital media, horror, reflexivity, and technologically-mediated performance.

Maria da Conceição Oliveira Guimarães é pós-doutorada em Literatura Estrangeira Moderna pela Universidade de Coimbra (CAPES), doutorada em Letras em Literatura e Cultura (UFPB com estágio doutoral na Universidade de Coimbra, CAPES), mestre em Literatura Comparada (UFRN), Specialist in Modern Masterpieces of World Literature (Harvard University, USA), investigadora da Universidade de Lisboa (projeto: «Portugueses de Papel» CLEPUL-FLUP), pesquisadora da IFRN (projeto: «Apocalípticas Integradas: Mulheres que tocam o terror», coordenadora do Clube de Leitura «Páginas Vívidas» na Biblioteca Estadual Câmara Cascudo (BECC) e membro das agremiações literárias «Mulherio das Letras Nísia Floresta» e «União Brasileira dos Escritores» (UBE/RN). Publicou *Matizes de uma poesia irisada: Sophia de Mello Breyner Andresen* (Ideia, 2015), *Antígona de Sófocles: uma leitura sob a «visão e paralaxe»* (Novas Edições Acadêmicas, 2017) e *O Canto helênico de Sophia em um tempo dividido* (Liber Ars, 2023). Venceu o «6º Prêmio AFEIGRAF de Poesia 2024» e destaques em Poesia Selo OFF-FLI em 2025 e Prêmio Terra em 2025.

Emily K. Johnson is an Associate Professor in the Department of English (Technical Communication and Digital Humanities), graduate faculty in the Technical Communication MA program, and core faculty in the Texts and Technology Ph.D. program at the University of Central Florida. She is co-author of *Critical Making in the Age of AI* (with Anastasia Salter, Amherst College P, 2025) and *Playful Pedagogy in the Pandemic: Pivoting to Games-Based Learning* (with Anastasia Salter, Routledge 2022). Johnson is Principal Investigator of a project funded by the U.S. Department of Education investigating the efficacy of five of her collaboratively designed video games for language learning. Her work has been published in *Technical Communication Quarterly, Communication Design Quarterly, Computers and Composition, Computers and Education,* the *Journal for Universal Computer Science,* and more.

Barry Mauer is associate professor of English at the University of Central Florida and author of *Deadly Delusions: Right-Wing Death Cult,* co-author of *Strategies for Conducting Literary Research* and *The Anti-Autocracy Handbook: A Scholars' Guide to Navigating Democratic Backsliding,* and co-editor of *Reimagining the Humanities.* His research focuses on cognitive immunity, which is the mind's ability to protect itself from pathological beliefs, and citizen curating, which brings ordinary people into the production of exhibits.

Guillem Molla is a Senior Lecturer of Spanish and Director of the Catalan Studies Program at the University of Massachusetts Amherst. He holds a Ph.D. in Spanish and Catalan Studies from the University of Girona and has published on comparative literature, literary journalism, documentary film, food anthropology, and reception studies. He is the author of *Ramon Esquerra: Geografia crítica d'un esperit comparatista* and editor of several critical volumes, including the correspondence between Josep Pla and Jaume Vicens Vives (*Ediciones Destino*) and a special double issue on contemporary Catalan literature in translation for *Metamorphoses*. He has taught literature, culture, and digital humanities in Spain and the UK, with appointments at the University of Bristol and Cardiff University. Founder of the Sant Jordi Translation Contest at UMass, he has been awarded research and teaching fellowships and actively contributes to cultural programming and curriculum development at the university and beyond.

Tiago Andreotti é Advogado e Professor da Universidade Federal de Mato Grosso do Sul. Concluiu o seu Doutorado em Direito pelo European University Institute e seu LL.M. pela New York University. Atua principalmente nas áreas de Direito Tributário, Direito Administrativo e Direito Contratual. Foi Gerente Jurídico da Companhia de Gás do Estado de Mato Grosso do Sul e Professor da Universidade Católica Dom Bosco. Está habilitado para a prática jurídica no Brasil e em Nova Iorque.

José Paulo Gutierrez. Professor da Faculdade de Direito da Universidade Federal de Mato Grosso do Sul (FADIR/UFMS). Graduado em Filosofia e Direito. Mestre em Direito e Economia. Doutor em Educação. Professor pesquisador e Vice-líder do Grupo de Pesquisa Antropologia, Direitos Humanos e Povos Tradicionais (UFMS). Tem experiência na área de Filosofia, Direito e Educação, com ênfase nas áreas de Direito Humanos, Antropologia e Sociologia Jurídica, Filosofia do Direito, Direitos Especiais, Educação em Direitos Humanos e Ética Profissional. Consultor Internacional voluntário para assuntos internacionais sobre Interseccionalidade e Direitos Humanos, Direitos Indígenas, Direitos Ambientais e Antropologia e Sociologia Jurídica no Instituto de Direitos Humanos de Mato Grosso do Sul José do Nascimento.

Ana Paula Martins Amaral é Professora titular da Universidade Federal de Mato Grosso do Sul. Professora da graduação e do Programa de Mestrado em Direito da Faculdade de Direito da Universidade Federal de Mato Grosso do Sul. Graduada em Direito pela Universidade Católica Dom Bosco (UCDB), Mestre e Doutora em Direito pela Pontifícia Universidade Católica de São Paulo (PUC/SP). Pós-doutorado pela Universidade Federal de Santa Catarina (UFSC). Autora de diversas obras publicadas em livros e revistas científicas. Pesquisadora, líder do grupo de pesquisa: Direito Internacional, Direitos Humanos e Relações Transfronteiriças.

Robert Simon serves as Professor of Spanish and Portuguese and Coordinator of the Spanish Program in the Department of World Languages and Cultures at Kennesaw State University. Included among his published works are his two most recent books: *The Purple Gladiolus and the Mystic's Map: Mystical Symbolism and the Posthuman in the 20th and 21st Century Poetry of Ana Rossetti* (2023), *From Post-Mortem to Post-Mystic: Blanca Andreu, Galicia, and the New Iberian Mysticism* (2019), along with over 40 critical articles, book chapters, and other studies that analyze Transnationalism and Mysticism in the contemporary poetry of Angola, Portugal, and Spain. He is also the author of ten collections of poetry, winner of the 2023 Radow College of Humanities and Social Sciences Outstanding Research and Creative Activities Award and the RCHSS 2024-2025 Conference USA Faculty Achievement Award at Kennesaw State.

Marcelo Pacheco Soares é Professor Titular do Instituto Federal de Educação, Ciência e Tecnologia do Rio de Janeiro. Doutor e Mestre em Literatura Portuguesa pela Universidade Federal do Rio de Janeiro, desenvolveu Pós-Doutorado em Estudos Literários na Universidade Federal Fluminense. Conduz pesquisas principalmente sobre Literaturas Fantásticas, sobretudo no gênero conto, com ênfase na obra de autores como José Saramago, Machado de Assis, Jorge de Sena e Teresa Veiga, tendo dezenas de artigos publicados no Brasil e no exterior acerca desses temas. No IFRJ, atuou como Conselheiro da área de Linguística, Letras e Artes no Conselho Acadêmico de

Pós-Graduação, Pesquisa e Inovação, Diretor de Apoio Técnico ao Ensino e Coordenador do Curso de Especialização em Ensino de Histórias e Culturas Africanas e Afro-Brasileiras, onde foi docente de Literaturas. Hoje, está cadastrado em três programas de Especialização: Estudos Linguísticos e Literários, Educação de Jovens e Adultos e Ensino de Matemática.

Sandra Sousa is associate professor in the Modern Languages and Literatures Department at the University of Central Florida, where she teaches Portuguese language, Lusophone Studies and Latin American Studies. Her research focuses on colonialism and post-colonialism, Portuguese colonial literature, race relations in Mozambique, war, dictatorship and violence in contemporary Portuguese and Luso-African literature, feminine writing in Portuguese, Brazilian and African literature. She is the author of *Ficções do Outro: Império, Raça e Subjectividade no Moçambique Colonial* (Esfera do Caos, 2015), *Portugal Segundo os Estados Unidos da América* (Theya Editores, 2021), and has co-edited *Visitas a João Paulo Borges Coelho. Leituras, Diálogos e Futuros* (Colibri, 2017), *The Africas in the World and the World in the Africas: African Literatures and Comparativism* (Quod Manet, 2022), and *Djaimilia Pereira de Almeida: Tecelã de mundos passados e presentes* (UMinho editora, 2023).

Mingyu Sun earned her Ph.D. in Computational Linguistics and master's degree in computer science from Michigan State University. She is currently the Director of the Language Resource Center at the University of Wisconsin-Milwaukee in the United States, and the President of the Midwest Association for Language Learning Technology. She also teaches Chinese language, Chinese calligraphy, and language and technology integration pedagogy. She conducts research in the areas of second language acquisition, teaching methodology, and technology enhanced language learning.

Gergana Vitanova is Professor at the University of Central Florida. Her research spans sociocultural issues in second language learning and teaching, and her primary area of investigation focuses on the identity and agency of second language teachers. She was particularly instrumental in bringing Bakhtin's dialogical framework to applied linguistics. She has published in various national and international peer-reviewed journals, such as *TESOL Quarterly, Journal of Language, Identity, and Education, Critical Inquiry in Language Studies, Language and Dialogue, System,* and others. She is also the author of the book *Authoring the Dialogical Self: Gender, Agency, and Language Practices* (John Benjamins, 2010) and a co-editor of several edited volumes, published by Multilingual Matters.

Aimee Schoonmaker is a Lecturer at the TESOL program at the University of Central Florida (UCF) in the United States. She teaches undergraduate and graduate level courses, among them Applied Linguistics, Grammar, Practicum, Listening & Pronunciation, and Computer Assisted Language Learning. Aimee holds a bachelor's degree in English and Portuguese, a master's degree in TESOL, and a Ph.D. in Education with a TESOL track. Her teaching and research interests revolve around Metalinguistic Awareness (MA), Corrective Feedback (CF) in writing, and more recently Artificial Intelligence in the EFL and ESL education realm. Prior to UCF, she worked in EFL settings in Brazil, predominantly in language institutes and K-12 English education.

www.ingramcontent.com/pod-product-compliance
Lightning Source LLC
Chambersburg PA
CBHW071625140626
46555CB00021B/57